ASTRAL
ASTRAL

AF522246

Ecobiology of Aquatic Insects

Ecobiology of Aquatic Insects

Editor

Professor (Dr.) Arvind Kumar

FLS (London), FASc (Swiss), FISEC, FSESc, FAZ, FZA (Gold Medalist)

Pro-vice Chancellor

S.K.M. University, Dumka – 814 101 (Jharkhand)

&

Dr. Harbhajan Kaur

Department of Zoology, Punjabi University, Patiala – 147 002

2008

DAYA PUBLISHING HOUSE

Delhi - 110 035

© 2008 ARVIND KUMAR (b. 1953 –)
HARBHAJAN KAUR (b. 1955 –)

ISBN10 81-7035-548-6
ISBN13 978-81-7035-548-9

All rights reserved. Including the right to translate or to reproduce this book or parts thereof except for brief quotations in critical reviews.

Published by : **Daya Publishing House**
1123/74, Deva Ram Park
Tri Nagar, Delhi - 110 035
Phone: 27383999
Fax: (011) 23260116
e-mail : dayabooks@vsnl.com
website : www.dayabooks.com

Showroom : 4760-61/23, Ansari Road, Darya Ganj,
New Delhi - 110 002
Phone: 23245578, 23244987

Laser Typesetting : **Classic Computer Services**
Delhi - 110 035

Printed at : **Chawla Offset Printers**
Delhi - 110 052

PRINTED IN INDIA

Preface

Aquatic organisms have preferred habitats, defined by physical, chemical and biological features. Variation in one or more of these can lead to stress on individuals and possibly a reduction in the total number of species or organisms that are present. In extreme situations of environmental change, certain species will be unable to tolerate the changes in their environment and will disappear completely from the area concerned, either as a result death or migration. Thus, the presence or absence of certain species or family groups or the total species number of abundance has been exploited as a means of measuring environmental degradation.

Aquatic insects are the biological variables for assessment of the structural and functional aspects of aquatic ecosystems. Through biomonitoring, the cumulative effect of all the pollutants can be determined and the overall health of the aquatic ecosystems could to properly assess. Aquatic insect is the best tool of biomonitoring which can be used as a cost-effective means for supplementing the physico-chemical techniques. In the present study, aquatic insects have been used to assess the pollution levels in the aquatic environment. Keeping in view the importance of aquatic insects as biotool, the present work has been undertaken.

This book is the compilation of esteemed articles of internationally acknowledged experts in the field of aquatic biology with the intention of providing a sufficient depth of the subject to satisfy the needs at a level which will be comprehensive and interesting. The present book will be useful to the students, research scholars, scientists in the field of Environmental management and ecoplanners, politicians and other people with similar interest.

My special thanks and appreciation go to the scientists whose contributions have enriched this volume. I wish to express my sincere gratitude to Dr. P. C. Hembram Hon'ble Vice Chancellor, S. K. M. University, Dumka, Professor M. C. Dash, Hon'ble former Vice Chancellor of Sambalpur University, Professor N. C. Datta of Calcutta University, Professor K. C. Pandey of Lucknow University, Professor S. K. Konar of Kalyani University, Professor D. K. Belsare of Bhopal University, Professor U. C. Goswami of Gauhati University, Professor P. C. Mishra of Sambalpur University, Professor P. Natarajan of Kerala University, Professor P. S. Murthy of Bangalore University, Professor Ajit Varma of JNU, Professor A. L. Bhatia of Jaipur University, Professor A. K. Mittal of BHU, Professor S. P. Hosmani of Mysore University, Professor K. Kapoor of Udaipur University, Professor K. B. Reddy of Nagarjuna University, Professor M. Vikram Reddy of Pondicherry University, Professor B. K. Tiwari of NEHU, Professor G. Tripathi of Jodhpur University, Professor R. Ramalingam of Annamalai University, Professor Sharif U. Ahmad of Nagaland University, Professor G. K. Kulkarni of Aurangabad University, Professor S. U. Meshram of Nagpur University, Professor S. K. Battish of PAU, Professor B. M. Sharma of Manipur University, Professor B. D. Joshi of Hardwar University, Professor G. C. Pandey of Faizabad University, Professor K. C. Sharma of Ajmer University, Professor M. Raziuddin of Hazaribag University, Professor U. S. Bagde of Mumbai University, Professor Gurdeep Singh of I. S. M., Dhanbad, Dr. P. K. Goel of Karad, Professor S. P. Roy and Shri Tribhuwan Poddar of Bhagalpur University, Bhagalpur for encouragements.

I also express my deep sense of gratitude to my parents whose blessings have always prompted me to pursue academic activities deeply. I am also thankful to my wife, *Kumari Bimla* and my two lovely sons, *Kumar Pallav Shivshankaran* and *Kumar Prasun Ramakrishnan* whose natural smiles extended to me relief all through this tiresome endeavour.

Last but not the least, I am also thankful to Mr. Anil Mittal, Proprietor, Daya Publishing House, New Delhi for taking keen interest in bringing out of this book. Finally, I will always remain a debtor to all my well-wishers for their blessings, without which this book would not have come into existence.

Professor A. Kumar

Harbhajan Kaur

Contents

Chapter 1

Quantitative Assessment of Predatory Insects and their Integration with Varied Factors in Some Fish Culture Ponds

Manish Chandra Varma[1], Shiv Kumar[1], Raghbendra Pratap[2] and Arvind Kumar[2]

[1]*University Department of Zoology, T.M. Bhagalpur University, Bhagalpur – 812 007*

[2]*Environmental Science Research Unit, Post Graduate Department of Zoology, S.K.M. University, Dumka – 814 101, India*

ABSTRACT

Faunistic composition, community structure and quantitative attributes of predatory aquatic insects have been investigated in two different fish culture ponds (Shahjangi and University managed pond) of Bhagalpur (Bihar) during the period of September 1999 to August 2001. Only three orders *viz.* Coleoptera, Hemiptera and Odonata have been found predaceous in habit. Out of 23 species of recorded insects only *Dineutes spinos* (Coleoptera), *Anispos sardea* (Homoptera) *Cordulegaster* sp. and *Mesoqumphus lineatus* (Odonata) were found most abundant and dominant species having maximum total annual mean abundance (N/m^2) (9.56 ± 2.72 ± 1.03 and 12.64 ± 2.92 ± 1.08) for *D. spinosus*: 11.00 ± 4.06 ± 1.17 and 21.46

± 9.2 ± 3.42 for *A. sardea;* 8.45 ± 2.05 ± 1.02 for *M. lineatus* and 12.24 ± 3.92 ± 2.46 for *Cordulegaster* sp., frequency index (0.266, 0.325 and 0.383, 0.350 for *D. spinosus;* 0.658, 0.683 and 0.741, 0.758 for *A. sardea;* 0.200, 0.191) for *M. lineatus* and 0.575, 0.533 for *Cordulegaster* sp. and percentage dominance (10.33; 12.23 per cent and 7.21; 7.69 per cent for *D.spinosus;* 27.84 ± 31.43 per cent and 17.96, 19.17 per cent for *A. sardea;* 7.01, 5.9 per cent for *M. lineatus* and 9.57, 9.69 per cent) for *Cordulegaster* sp. in both the ponds. The significant correlation was also established between the physico-chemical constituents and total annual mean insect abundance. The main factors influencing the predatory insect population are apparently complex with complex extrinsic and intrinsic physical, chemical and biological factors.

Keywords: *Abundance, Frequency index, Percentage dominance, Aquatic insects.*

Introduction

Freshwater systems harbour diverse types of heterotropic communities specially the zooplankton, macro-invertebrates and fishes (Varma, 1990). They also act as an indicator of trophic structure, water quality and eutrophication of the aquatic ecosystem (Varma and Pratap, 2006).

Among the invertebrate taxa, predatory aquatic insects form an important component of the food chains and energy flow pathways and comprise a high proportion of biomass in freshwater. The predatory insects associated with the lentic and lotic systems usually belong to orders Coleoptera, Hemiptera and Odonata and are predaceous in habit. Literature survey reveals that extensive studies have been conducted on distribution, composition, systematics of these insects and their destructive role in fish culture ponds in India and abroad (Pakrasi, 1956; Tonapi and Ozarkar, 1969; Bose and Sen, 1978; Roy and Munshi, 1978; Roy, 1982; Sinha, 1992). However, except the preliminary report on seasonal variations in biomass and production of aquatic insects in some fish culture ponds of Bhagalpur by Varma and Kumar (2006), no information is available on the quantitative attributes of predatory insects in various lentic systems of Bhagalpur. Therefore, in the present study, an attempt has been made to evaluate the distribution, seasonal occurrence and abundance of predatory insects and their interaction with abiotic factors in two different fish culture ponds at Bhagalpur (Bihar) having latitude 25°15′ N and longitude 87°02′ E.

Materials and Methods

In the present study, two fish ponds were selected *viz.*, a University managed pond and Shahjangi Talab, a derelict pond situated in Bhagalpur township. The investigation was conducted from September 1999 to August 2001. Monthly random sampling of insect associated with the roots of free-floating macrophytic communities in the littoral and benthic zones of the selected water bodies were done by specially designed 1.6 m^2 lift net made of nylon cloth (mesh size 40 to 80 cm^{-2}). The collection of predatory insects was done at five sites according to the method suggested by Merrit and Cummins (1978) after slight modification. The total seasonal abundance of the insect population was determined monthly from the 10 samples taken from five sites for each selected water body. Insects of each sampling were sorted out species-wise and were preserved in 5 per cent formalin. Finally, insects of each sampling unit were counted and identified. The physical and chemical characteristics of water samples of both the ponds were analysed by following standard method of Welch (1952) and APHA (1981). The quantitative attributes of these insects are total annual mean abundance, frequency index and percentage dominance.

Abundance

The abundance represents the numerical strength of a species in the community and is expressed as a mean/average of the number of individual of any species per sampling unit of occurrence in a particular area of the habitat. The total annual numerical mean abundance (N/m^2) of all the sampled insect species were calculated by simply counting their number in all sampling units.

Frequency Index

The micro-distribution pattern of many population of animals is such that any given species may be encountered in some of the samples but not in others, and hence has a frequency of occurrence. The relative frequency of occurrence for all the species was determined by calculating their frequency indices by the formula (Raunkiaer, 1934):

$$\text{Frequency index} = \frac{\text{Number of sampling unit in which species is present (a)}}{\text{Total Number of samples examined (n)}}$$

Dominance

The percentage dominance of predatory aquatic insect species was calculated using the following formula (Wallwork, 1976):

$$\text{Dominance (per cent)} = a/n \times 100$$

where, 'a' is the number of individual species in all the sampling units examined, and 'n' is the total number of individuals of all the species examined in all sampling units.

Results

The predatory aquatic insects collected from two different stagnant water bodies (University managed and a derelict Shahjangi Talab pond) belong to three insect orders *viz.* Coleoptera, Hemiptera and Odonata. In case of Coleoptera and Hemiptera both larvae/ nymphs and adults were present in the sample where as in case of Odonata, only nymphs were collected. In the present study a total of three genera (*Cybister* sp., *Laccophilus* sp. (Fam-Dytiscidae), *Dineutes* sp. (Fam-Gyrinidae) of order Coleoptera; six genera *Anisons* sp. (Fam-Notonectidae), Diplonychus sp., *Sphaerodeama* sp. (Fam-Belostomatidae), *Plea* sp. (Fam-Pleidae), *Ranatra* sp., *Laccotrephes* sp. (Fam-Nepidae) of order Hemiptera and finally six genera, of order Odonata; *Mesogomphus* sp. (Fam-Gomphidae), *Cordulegaster* sp. (Fam-Cordulegasteridae), *Potamarcha* sp., Zyxomma sp. (Fam-Libellulidae), *Ischnura* sp., Agrionemis sp. (Fam-Coenagrionidae) were reported in almost all the sampling units.

The numerical total annual mean abundance, frequency index and percentage dominance of sampled insect species in both water bodies are summarized in Tables 1.1–1.6.

The result of total annual numerical mean abundance (N/m^2) in all sampled insect species reveals that among the members of

Coleoptera, *Dineutes spinosus* has higher values in both University managed and derelict ponds (9.56 ± 2.72 ± 1.03; 8.75 ± 2.80 ± 0.80 and 12.64 ± 2.92 ± 1.08; 16.26 ± 8.50 ± 2.64) for the year 1999–01 and 2000–01 respectively. In Hemiptera, *Anisops Sardea* is considered the most abundant aquatic insect species because the values of total annual mean abundance in two lentic water bodies were 10.5 ± 4.64 ± 1.33; 11.00 ± 4.06 ± 1.17 in University managed pond and 14.68 ± 4.98 ± 2.42 ± 9.21 ± 3.42 in Shahjangi Talab derelict pond during the 1999-2001. Similarly, among the Odonates, *Mesogomphus lineatus* has the higher values of total annual mean abundance (6.25 ± 3.01 ± 0.92; 8.45 ± 2.05 ± 1.02) in University managed pond whereas in Shahjangi Talab derelict and *Cordulegaster* Sp. has higher values (11.24 ± 2.42 ± 1.74; 12.24 ± 3.92 ± 2.46) during the period 1999-2001.

Table 1.1: Total Annual (mean ± SD; ± SE) Abundance (N/m^2) of All Sampled Predatory Insect Species of University Managed Fish Pond at Bhagalpur during the Period 1999–01

Order/Species	*1999-00 Total Annual Mean Abundance/sp.*	*2000-01 Total Annual Mean Abundance/sp.*
	Coleoptera	
Cybister confusus	2.01 ± 1.66 ± 0.58	1.92 ± 1.52 ± 0.55
C. regulosus	1.46 ± 1.16 ± 0.42	1.02 ± 0.83 ± 0.29
Leccophilus anticatus	6.42 ± 3.37 ± 0.97	4.06 ± 1.92 ± 0.41
Dineutes spinosus	9.56 ± 2.72 ± 1.03	8.75 ± 2.80 ± 0.80
	Hemiptera	
Anisops sarden	10.5 ± 4.62 ± 1.33	11.00 ± 4.06 ± 1.17
Ranatra filiformis	3.33 ± 2.46 ± 0.7	3.58 ± 2.64 ± 0.76
R. elougata	2.18 ± 1.66 ± 0.46	1.82 ± 1.33 ± 0.52
Laccotrephas rubber	4.25 ± 3.81 ± 1.10	1.82 ± 1.33 ± 0.52
Diplonychus annulatum	7.0 ± 2.91 ± 1.02	7.40 ± 2.62 ± 1.21
	Odonata	
Mfesagomphus lineatus	6.25 ± 3.01 ± 0.92	8.45 ± 2.05 ± 1.02
Potamarcha abscura	4.33 ± 3.86 ± 1.11	4.20 ± 2.04 ± 0.84
Ischnura delicata	5.23 ± 2.86 ± 1.21	4.92 ± 1.42 ± 1.23
Agriochenis sp.	2.42 ± 1.24 ± 0.4	3.51 ± 1.92 ± 0.76

Table 1.2: Total Annual (mean ± SD; ± SE) Abundance (N/m²) of All Sampled Predatory Insects Species of Shahjangi (Derelict) Fish Pond at Bhagalpur during the Period 1999–01

Order/Species	*1999-00 Total Annual Mean Abundance/sp.*	*2000-01 Total Annual Mean Abundance/sp.*
	Coleoptera	
Cybister confusus	2.5 ± 0.92 ± 0.24	3.2 ± 0.86 ± 0.36
C. limbatus	1.89 ± 0.62 ± 0.16	2.2 ± 0.72 ± 0.29
C. tripunctatus asiaticus	1.05 ± 0.75 ± 0.21	2.5 ± 0.81 ± 0.24
Laccohilus parvulus	4.41 ± 1.62 ± 0.89	5.2 ± 1.92 ± 0.89
Dinsutes spinosus	12.64 ± 2.92 ± 1.08	16.26 ± 8.50 ± 2.64
	Hemiptera	
Anisops sardea	14.68 ± 4.98 ± 2.42	21.46 ± 9.21 ± 3.42
A. breddinni	7.21 ± 2.25 ± 1.69	6.98 ± 2.64 ± 1.46
Plea frontalis	6.27 ± 2.08 ± 0.98	5.89 ± 1.68 ± 1.46
Ranatra elongata	2.38 ± 0.72 ± 0.68	3.58 ± 1.26 ± 0.74
Laccotrephes griseus	8.46 ± 2.46 ± 1.08	9.24 ± 1.42 ± 0.69
Diplonychus annulatum	11.64 ± 4.26 ± 1.92	12.64 ± 3.64 ± 2.42
Sphaarodema rusticum	3.65 ± 2.56 ± 0.92	4.52 ± 2.68 ± 0.92
	Odonata	
Mesogamphus lineatus	8.54 ± 2.92 ± 1.02	9.21 ± 3.46 ± 1.69
Cordulegaster sp.	11.24 ± 2.42 ± 1.74	12.4 ± 3.92 ± 2.46
Ischnura delicata	7.56 ± 3.24 ± 1.64	8.24 ± 2.91 ± 1.22
I. senegalensis	4.46 ± 1.84 ± 0.55	3.85 ± 1.24 ± 0.46
Zyxomma netiolatum	7.00 ± 1.02 ± 1.21	8.56 ± 1.46 ± 0.51
Potamarcha obscura	5.46 ± 1.46 ± 0.82	6.42 ± 2.42 ± 1.02

The result of frequency index (FI) of all studied insect species revealed that among the Coleoptera, *Dineutes Spinosus* has higher FI (0.266; 0.325) followed by *Laccophilus anticatus* (0.175; 0.158) in University managed pond during the period 1989–91. In Shahjangi Talab derelict pond, it was again *Dineutes spinosus* which has higher frequency presence in the sampling unit (46 times in 1999–2000 and 42 times in 2000–01 respectively). Among the Hemiptera, *Apisops sardea* has maximum number of appearance in the sampling units (79 times in 1999–2000) and 82 times in 2000–01 for University managed pond and 89 times 1999-01 and 91 times in 2000-01 for Shahjangi Talab derelict pond. The Frequency Index (FI) of *A. sardea* were recorded 0.658 and 0.683 University managed pond and 0.741 and 0.758 for Shahjangi Talab pond. Among the Odonates, *M. lineatus* and *Ischnura delicata* has higher FI values (0.200; 0.191 and 0.175; 0.158) in University managed pond during the period 1999–2000, while in Shahjangi Talab pond, it was *Cordulegaster* sp. which has maximum FI values (0.575; 0.533) followed by *Potamarcha obscura* (0.283; 0.300) during the period 1999–01.

Table 1.3: Frequency Index of All Sampled Predatory Insect Species of University Managed Fish Pond at Bhagalpur During the Period 1999–01

Name of the Species	*No. of Sampling Units in which Species is Present (a)*		*Frequency Index*	
	1999–00	*2000–01*	*1999–00*	*2000–01*
Cybister confusus	18	16	0.150	0.133
C. regulosus	9	7	0.075	0.058
Laccophilus anticatus	21	19	0.175	0.158
Dineutes spinosus	32	39	0.266	0.325
Anisops sardea	79	82	0.658	0.683
Ranatra filiformis	12	9	0.100	0.075
R. elongata	8	7	0.066	0.058
Laccotrephes rubber	19	17	0.158	0.141
Diplonychus annulatum	26	29	0.216	0.241
Mesoqomphus lineatus	24	23	0.200	0.191
Potamarcha obscura	18	12	0.150	0.100
Ishnura delicata	21	19	0.175	0.158
Agriocnenis sp.	20	17	0.166	0.141

Total Number of samples examined (n) 1999–00 = 120; 2000–01 = 120.

Table 1.4: Frequency Index of All Sampled Predatory Insect Species of Derelict Pond Shahjangi Fish Pond at Bhagalpur During the Period 1999–01

Name of the Species	No. of Sampling Units in which Species is Present (a)		Frequency Index	
	1999–00	2000–01	1999–00	2000–01
Cybister confusus	21	24	0.175	0.200
C. limbatus	8	6	0.066	0.050
C. tripunctatus asiaticus	7	5	0.058	0.041
Laccophilus anticatus	32	28	0.266	0.316
Dineutes spinosus	46	42	0.383	0.350
Anisops sardea	89	91	0.741	0.758
A. breddinni	26	28	0.216	0.233
Plea frontalis	23	19	0.191	0.158
Ranatra filiformis	9	11	0.075	0.091
Laccotrephes rubber	39	41	0.325	0.341
Diplonychus annulatum	46	51	0.383	0.425
Mesoqomphus lineatus	34	33	0.283	0.275
Cordulegaster sp.	69	64	0.575	0.533
Ischnura delicata	30	28	0.250	0.233
Zyxomma petiolatum	23	22	0.191	0.183
Potamarcha obscura	18	12	0.150	0.100

Total Number of samples examined (n) 1999–00 = 120; 2000–01 = 120.

The result of percentage dominance of all studied insect species revealed that in University managed pond, *Dineutes spinosus* (Coleoptera) has maximum values of Dominance (10.33; 12.23 per cent) followed by *Laccophilus anticatus* (8.01; 6.75 per cent) during the both years of study. In Shahjangi Talab pond, it was again *D. spinosus* which has higher values of percentage dominance (7.21; 7.369 per cent). Among the Hemiptera, *A. sardea* and *D. annulatum* has maximum percentage dominance values (27.84; 31.43 per cent and 8.43 per cent; 10.97 per cent) in University managed pond during the two years of study. *Anisops* sp. and *Dilonychus* sp. were again found dominant in Shahjangi Talab pond having higher values of percentage dominance (17.46; 19.17 per cent and 8.5; 8.02 per cent) in both the years. Among the Odonates, *M. lineatus* and *Ischnura delicata* have higher values of dominance (8.01; 5.90 per cent and 6.75; 5.48 per cent) in University managed pond. But in derelict pond, *Cordulagaster* sp. has higher values of dominance (9.57; 6.69 per cent) during the two years of study.

Table 1.5: Percentage Dominance of All Sampled Predatory Insect Species of Shahjangi Managed Fish Pond at Bhagalpur During the Period 1999–01

Name of the Species	*No. of Sampling Units in which Species is Present (a)*		*Frequency Index*	
	1999–00	*2000–01*	*1999–00*	*2000–01*
Cybister confusus	22	26	4.64	5.48
C. regulosus	16	12	3.37	2.53
Laccophilus anticatus	38	32	8.01	6.75
Dineutes spinosus	49	58	10.33	12.23
Anisops sardea	132	149	27.84	31.43
Ranatra filiformis	18	12	3.79	2.53
R. elongata	14	10	2.95	2.10
Laccotrephes rubber	30	28	6.32	5.90
Diplonychus annulatum	40	52	8.43	10.97
Mesoqomphus lineatus	38	28	8.01	5.90
Potamarcha obscura	20	16	4.21	3.37
Ishnura delicata	32	26	6.75	5.48
Agriocnenis sp.	25	21	5.27	4.43

Total Number of samples examined (n) 1999–00 = 474; 2000–01 = 470.

The concentration of physico-chemical constituents in these two different ponds water was estimated and the correlation coefficient ('r' values) between these physico-chemical factors (water temp. pH, DO_2, FCO_2; Chloride and phosphate) with total annual mean predatory insects abundance (N/m^2) in these wetlands of Bhagalpur have been computed and shown in Table 1.7.

Table 1.6: Percentage Dominance of All Sampled Predatory Insect Species of Derelict Pond Shahjangi Talab Fish Pond at Bhagalpur During the Period 1999–01

Name of the Species	*No. of Sampling Units in which Species is Present (a)*		*Frequency Index*	
	1999–00	*2000–01*	*1999–00*	*2000–01*
Cybister confusus	26	30	3.07	3.34
C. limbatus	14	9	1.65	1.00
C. tripunctatus asiaticus	11	10	1.30	1.11
Laccophilus anticatus	49	51	5.79	6.02
Dineutes spinosus	61	69	7.21	7.69
Anisops sardea	152	172	17.96	19:17
A. breddinni	41	49	4.84	5.46
Plea frontalis	39	28	4.60	3.12
Ranatra filiformis	13	16	1.53	1.78
Laccotrephes rubber	56	61	6.61	6.80
Diplonychus annulatum	69	72	8.15	8.02
Sphaerodema rusticum	19	21	2.24	2.34
Mesoqomphus lineatus	56	66	6.61	7.35
Cordulegaster sp.	81	87	9.57	9.69
Ischnura delicata	32	29	3.78	3.23
I. senegalensis	42	31	4.96	3.45
Zyxomma petiolatum	34	37	4.01	4.12
Potamarcha obscura	51	56	6.02	6.24

Total Number of samples examined (n) 1999–00 = 846; 2000–01 = 897.

In the present study, a significant correlation of temperature ($r = 0.58$; 0.56, $P < 0.10$ in University managed pond, $r = 0.68$; $P < 0.05$, 0.44 in Shahjangi Talab pond); pH ($r = 0.64$, $P < 0.66$, $P < 0.05$ in University managed $r = 0.73$, 0.70; $P < 0.01$ in derelict pond); DO_2 (0.62, 0.49; $P < 0.10$ in University managed, $r = 0.68$, $P < 0.05$ in Shahjangi Talab pond; and Chloride ($r = 0.60$, 0.66, $P < 0.05$ in University managed and $r = 0.59$, 0.57, $P < 0.10$ in derelict pond) with total annual insect abundance in these two water bodies have been established. The other abiotic factors such as FCO_2 and phosphate were found negatively correlated with total annual insect abundance values in University managed pond (r = –0.55, –0.45; P < 0.10 for FCO_2 and r = 0.0244, 0.68 for phosphate), while in Shahjangi Talab pond, both FCO_2 and phosphate were again found negatively correlated (–0.68, 0.63, P < 0.05 for FCO_2 and –0.68; –0.57, P < 0.05 for phosphate).

Table 1.7: Correlation Coefficient Values of Total Annual Insect Abundance with Physico-chemical Constituents in a University Managed and a Derelict Pond During the Period 1999–01

Parameters	*Coefficients of Correlation (r) 1999–00*	*Student 't' Test (t) 2000–01*	*Coefficient Correlation (r)*	*Student 't' Test (t)*
	In a Govt. Managed Fish Pond			
Water Temp.	0.5823	2.0786*	0.5610	2.031*
pH	0.6434	2.742**	0.6641	2.836
Dissolved Oxygen	0.6231	2.412*	0.4923	1.864
Free CO_2	–0.5507	2.026*	0.4509	1.597
Chloride	0.5042	2.421*	0.6680	2.852**
Phosphate	–0.0244	0.0774	–0.876	0.2781
	In a Shahjangi Fish Pond			
Water Temp	0.6831	2.9231**	0.4472	0.7210
pH	0.7327	3.404**	0.7099	3.187***
Dissolved Oxygen	0.6837	2.954**	0.3933	1.353
Free CO_2	–0.6988	3.005**	–0.6301	2.566**
Chloride	0.5944	2.3140*	0.5712	2.242*
Phosphate	–0.6847	2.970**	–0.5712	2.263 *

*: Significant at 10 per cent ($p < 0.10$); **: Significant at 5 per cent ($p < 0.05$); ***: Significant at 1 per cent ($p < 0.01$).

Discussion

In the understandings of quantitative attributes of the community, a distinction has to be made between abundance, frequency and dominance of insects. Abundance includes the number of animals of a unit at given time. So it is a more dynamic idea which reveals that a population is always changing *i.e.* increasing or decreasing at any instant of time and then to quantify its size at any one moment. It is time relative parameter of the community. On the other hand, frequency is the number of occurrence of the species in a sample because some species may be present in a sample or absent in another. An index expressing the degree of occurrence is essential in order to know the composition of the community in any environment.

Natural communities have large number of species but only a few species are generally exerting the major controlling influences on the community by virtue of their number, size, biomass, productivity, turn-over rates and other activities. Such species are said to be dominant and play a major role in determining the structure and function of the entire community. Therefore, the analysis of community structure of any aquatic heterotrophs, such quantitative traits as abundance, frequency and dominance will be the essential parameters.

In the present investigation, the two years observations on the total annual mean abundance, frequency index and percentage dominance of predatory insects in two different types of freshwater habitats of Bhagalpur showed some interesting fluctuation in their population. Such fluctuations in their population were also recorded by various workers in the past. In all studied water bodies of Bhagalpur the member belonging to orders Coleoptera, Hemiptera and Odonata were recorded *i.e. Cybister confusus, Laccophilus* sp., *Denutesspinusus, Anisops sardea, Diplonychus annulatum, Laccotrephes* sp., *Mesogomphus lineatus, Cordulegaster* sp., *Potamarcha obscura* and *Ischnura delicata* and may be considered as well represented forms contributing substantially to the predatory aquatic insect fauna showing definite trends in their seasonal population fluctuation. This is obvious from the result of total annual mean abundance (N/m^2) of predatory insect recorded from both water bodies. In Coleoptera, *Dineutes spinosus* has higher values of it *i.e.* 9.56 ± 2.72 ± 1.03; 8.75 ± 2.80 and 12.64 ± 2.92 ± 1.08; 16.26 ± 8.50 ± 2.64 in both ponds during the period 1999–01. Among Hemiptera, *Anisops sardea* has maximum total annual mean abundance in these two types of water bodies (10.5 ± 4.64 ± 1.33; 11.00 ± 4.06 ± 1.17 and 14.68 ± 4.98 ± 2.42; 21.46 ± 9.21 ± 3.42) during the period 1999–01. Similarly among the Odonates, *Mesogomphus lineatus* and *Cordulegaster* sp. were the most abundant species in both the ponds having maximum values of total annual mean abundance (6.25 ± 3.01 ± 0.92; 8.45 ±

2.05 ± 1.02 for *M. lineatus* and 11.24 ± 2.42 ± 1.74; 12.24 ± 3.92 ± 2.46 for *Cordulegaster* sp.).

The result of frequency index and percentage dominance of these recorded species revealed that among the Coleoptera, *D. spinosus* and *Laccophilus* sp. were the dominant species having higher values of FI and percentage dominance. In Hemiptera, *Anisops sardea* was the most dominant in both the ponds having higher frequency indices (0.658; 0.683 and 0.741; 0.758) and percentage dominance (27.84; 31.43 per cent and 17.96; 19.17 per cent) during the period 1999–01. Similarly among the Odonates, *M. lineatus* and *Cordulegaster* sp. were the dominant species, because of their higher FI and percentage dominance values. Similar trends of result was also reported by Julka (1977) in a fish pond at Barrackpore (W.B.).

Certain physico-chemical constituents such as temperture, pH, DO_2, FCO_2 chloride and phosphate are the most apparent abiotic factors which affect the various quantitative attributes of aquatic insects in natural lentic system. This is obvious from the significant correlation values (r) of total annual insect abundance with some important physico-chemical constituents in these two water bodies during the period 1999–01. The significant positive correlation was found between total annual insect abundance and water temperature ($r = 0.58, 0.56, P < 0.10$, in University managed and $r = 0.68, 0.44, P < 0.05$ in derelict pond); pH ($r = 0.64, 0.66, P < 0.05$ for University managed pond and 0.78, 0.70, $P < 0.01$ in derelict pond); DO_2 ($r = 0.62, 0.49, P < 0.10$ in University managed pond and $r = 0.68$, 0.39, $P < 0.05$ in derelict pond) and chloride ($r = 0.60, 0.66, P < 0.05$ in University managed pond and $r = 0.59, 0.57, P < 0.10$ in Shahjangi Talab pond). However, free CO_2 and phosphate were found negatively correlated with total annual insect abundance values in both the ponds.

It is interesting to find that the derelict pond has larger numbers of insect species than in what the University managed pond has. This might be due to the abundance of macrophytic vegetations in the littoral zones of the derelict pond. These macrophytes provide an excellent diverse niches for several insect larvae/adults which are adapted for mining into stems and leaves and also provide suitable substrate for adult stages of insects. But in the managed pond, there is a regular deweeding programme in which removal of noxious weeds takes place. The absence of macrophytes in the managed pond might be discouraging the growth and colonization of predatory aquatic insects.

Thus, the result of total annual numerical mean abundance, frequency index and percentage dominance of studied predatory aquatic insect species in these two types of freshwater bodies of Bhagalpur showed that those species which have higher values of abundance in a particular habitat are not considered as dominant species. The dominant species are those which have maximum total annual mean abundance (N/m^2) along with higher frequency index values *i.e.* maximum appearance in the sampling units.

References

APHA, 1981. *Standard Methods for Examination of Water and Wastewater*, 15th edn. American Public Health Association, Washington, D.C.

Bose, K.C. and Sen, N.S., 1978. *Geobios*, 5(5): 212–213.

Biggs, H.J.F., 1982. *New Zealand Journal of Marine and Freshwater Research*, 16: 81–88.

Das, S.M. and Bisht, R.S., 1979. *Ind. J. Ecol.*, 5(1): 35–40.

Fernando, C.H. and Galbraith, D., 1973. *Varh. Intenat. Verein. Limnol.*, (18): 1564–1575.

Hafiz, H.A. and Mathai, G., 1938. *Rec. Ind. Mus.*, 2: 207–210.

Julka, J.M., 1977. *Oriental Insects*, 11(1): 139–149.

Meritt, R.W. and Kenneth, W. Cummins (Eds.), 1978. *An Introduction to the Aquatic Insects of North America*. Kendall/Hunt Publishing Co., 2460, Kerper Boulevard, Dubugue, Iowa, 520011, U.S.A., pp. 427.

Pakrasi, B.B., 1956. *Proc. Ind. Acad. Sci.*, 38B(5): 211–213.

Pruthi, H.S., 1923. *Int. Revue gas. Hydrobio.*, 29: 46–47.

Rao, T.K.R., 1977. *Entomon.*, 1: 123–132.

Raunkaier, C., 1934. *The Life Form of Plants and Statistical Plant Geography*. Oxford Clarendon Press.

Roy, S.P., Kumar, V. and Pathank, H.S., 1988. *Rec. Zool. Suv., India*, 85(1): 49–57.

Roy, S.P. and Munshi, J.S., 1979. *The Indian Journal of Zootomy*, **20(3)**: 139–146.

Roy, S.P., 1982. *Oriental Insects*, 16(1): 55–62.

Sinha, K.K. and Sinha, D.K., 1990. *Environment and Ecology*, 8(4): 1231–1234.

Sinha, D.K. and Roy, S.P., 1991. *J. Freshwater Biol.*, 3(1): 99–103.

Sinha, D.K., 1992. *Ph.D. Thesis*, Bhagalpur University, India, pp. 35–54.

Tonapi, G.T., 1922. *Proc. Nat. Inst. Sci., India*, 25: 321–332.

Tonapi, G.T. and Ozarkar, V.A., 1969. *J. Bombay Nat. Hist. Soc.*, 66(2): 310–316.

Varma, M.C., 1990. *Ph.D. Thesis*, T.M. Bhagalpur University.

Varma, M.C. and Pratap, Raghbendra, 2006. *J. Haematol. and Ecotoxicol.*, 1(1): 13–20.

Varma, M.C. and Kumar, Shiv, 2006. *J. Haematol. and Ecotoxicol.*, 1(1): 57–62.

Wilson, C.B., 1923. *Bulletin of the Bureau of Fisheries*, 30: 9–38.

Weilderholm, T., 1980. *J. Water Poll. Cont. Fed.*, 52: 537–547.

Wallwork, J.A., 1976. *The Distribution and Diversity of Soil Fauna.* Academic Press, London, pp. 335.

Welch, P.S., 1952. *Limnological Methods.* Mc-Graw Hill Book Company, Inc., New York, London.

Chapter 2

Genotoxicity of Cadmium Assessed by the Micronucleus Test on *Channa punctatus* (Pisces : Family–Channidae)

N.K. Tripathi and Anurag Dubey
P.G. Department of Zoology, University of Jammu, Jammu – 180 006 (J&K)

ABSTRACT

The genotoxicity of the cadmium chloride was studied using the micronucleus (MN) test in the kidney cells of *Channa punctatus*. The frequency of micronuclei was examined from fishes exposed to two different concentrations (1 mg/l and 2 mg/l) of the compound and sacrificed at four sampling times (24, 48, 72 and 96 hrs). Results obtained demonstrated the genotoxic effects of the cadmium in the experimental model employed. The micronuclei were detected in treated and untreated kidney cells. However, the frequency of micronuclei was significantly higher in the treated fishes as compared to that of the control fishes. The increase in micronucleated cells (MNC) was up to 96 hrs when treated with 1 mg/l; up to 72 hrs when treated with 2 mg/l and after 96 hrs a decline in number of MNC was observed with 2 mg/l.

Keywords: *Genotoxicity, Cadmium chloride, Micronucleus test, Kidney cells.*

Introduction

As a result of industrial activities the aquatic ecosystem has been increasingly contaminated by heavy metals. This poses great threat to aquatic organisms in particular and to the whole ecosystem in general (Zou and Bu, 1994). In this respect cadmium pollution is of global concern (Samiullah, 1990). Cadmium finds its way via atmospheric transport into the environment from a number of industries. It is a by-product of fossil fuel combustion and base metal smelting. This biologically non-essential heavy metal is highly toxic to aquatic organisms (Rishi and Jain, 1998). The harmful effects of cadmium is attributed to its effects on sulfhydryl groups of enzymes especially dehydrogenases (Belies, 1975). Moreover, mitochondria are key intracellular targets for cadmium (Sokolova, 2004).

In recent years, fishes have received special attention as possible monitors of environmental pollution. Fishes are excellent model system for the study of the mutagenic and/or carcinogenic potential of contaminants present in water sample since they can metabolize, concentrate and store water borne pollutants (Al-Sabti, 1991). Since fish often respond to toxicants in a similar way as do higher vertebrates, so they can be used to screen for chemicals that are potentially teratogenic and carcinogenic in humans. The main application for model systems using fish is to determine the distribution and effects of chemical contaminants in the aquatic environment (Al-Sabti and Metcalfe, 1995).

Several tests have been developed with aquatic animals which can potentially be used to assess the genotoxicity of chemicals, using such endpoints as chromosomal aberrations, sister-chromatid exchanges and micronuclei (MN) (Hooftman, 1981; Alink *et al.*, 1980 and Al-Sabti, 1986). The MN assay has been used as a measure of genotoxicity in fish under laboratory and field conditions (Das and Nanda, 1986; Hooftman and de Raat, 1982 and Matsumoto *et al.*, 2003) and has shown potential for *in situ* monitoring of water quality.

MN are small fragments of intracytoplasmic chromatin which arise from chromosome breaks or whole chromosomes, after the action of clastogenic substances or spindle-poisons, that do not migrate during anaphase (Al-Sabti and Metcalfe, 1995).

The fresh water fish *Channa punctatus* was chosen to evaluate the genotoxic potential of the cadmium chloride by using the MN test.

Materials and Methods

Fish

The choice of *C. punctatus* was based on the fact that it is a common and relatively stationary species with good availability in local fresh water sources in the Jammu district of Jammu and Kashmir state, India. In addition, it is relatively easy to handle and acclimatize to laboratory condition because of presence of accessory air-breathing organs.

All the adult fish specimens (12–15 cm in length and 30–60 gm in weight) used in this experiment were collected from local fresh water sources in Jammu district. Before the experiment, fishes were acclimatized in aquaria containing well-aerated, dechlorinated tap water at room temperature and a constant photoperiod (12 hr light/ dark cycles). The fishes were fed fish food daily during holding periods.

Chemicals

The studied compound was cadmium chloride ($CdCl_{2.5}H_2O$) which was used in the present experiment.

Experimental Design

Tests were conducted in dechlorinated tap water and in accordance with standard methods. The exposure system consisted of well-aerated plastic containers in which the test metal salt solution was added.

For each concentration and duration, a group of 4 fishes were exposed to test solution of $CdCl_2$ in standard water at room temperature. The sub-lethal concentration of $CdCl_2$ tested were 1 mg/l and 2 mg/l for 24, 48, 72 and 96 hrs each. The test concentrations were renewed thrice a week. Controls were also run simultaneously and 2 fishes as a control parallel to each concentration was used.

Slide Preparation

Fishes were dissected to remove the anterior portion of kidneys and were treated in 0.056 per cent KCl hypotonic solution and kept as such for 25–30 minutes. It was then fixed in 1 : 3 ratio of glacial acetic acid and methanol. Several changes of fixative were given. Then the prepared slides were stained with dilute solution of 2 per cent Giemsa stain. The slides were differentiated in distilled water and then air-dried. The slides were scored using a Nikon microscope using 1000x magnification. The frequency of micronucleated cells (MNC) per 1000 cells was determined for each fish.

Results

The cadmium chloride treatment caused damage to the nucleus as revealed by the frequency of micronucleated cells determined in the different treatments and durations (Table 2.1).

Table 2.1: Frequency of Micronucleated Cells (MNC) in Kidney Cells of *C. punctatus* Treated with Cadmium Chloride at Different Doses in Relation to Duration of Exposure

Concentration mg/l	*Duration in hrs.*	*No. of Specimens*	*No. of Cells Scored*	*Total MNC*	*MNC per 1000 Cells Observed (mean±SD)*
1.0	Control	2	2600	7	2.69±0.26
	24	4	4439	23	5.16±0.11
	48	4	5918	45	7.62±0.25
	72	4	5721	84	14.8±0.19
	96	4	4417	81	18.35 ± 0.09
2.0	Control	2	2945	6	2.0 ± 0.42
	24	4	5639	59	10.42 ± 0.3
	48	4	5135	73	14.17 ± 0.27
	72	4	6309	111	17.6 ± 0.2
	96	4	5546	83	15.1 ± 0.08

A single MN was predominant in the kidney cells analysed. Only few of all counted MNCs had two MN. The size as well as location of MN within the cytoplasm varied from cell to cell; the shape was generally round in almost all the cells. In addition to the incidence of MN, a few cells exhibited some nuclear anomalies *e.g.* some cells contained two normal nuclei; in some cells the MN was connected to the main nucleus by a chromatin bridge. These nuclear anomalies were not noted in the controls.

There was elevated response in the incidence of MNCs co-related with the increase in the duration of exposure (24 to 96 hrs) and in both the concentrations tested (1 mg/l and 2 mg/l) except at the concentration of 2 mg/l after 96 hr, it was reduced. This elevated response in the MNC incidence differed from parallel controls.

Discussion

In the present study it has been observed that *C. punctatus* reacted to the exposure of cadmium taken as cadmium chloride through changes in the structural organisation of the nucleus (*i.e.* by the formation of MN) suggesting a reaction to cadmium stress.

The MN formation presently reported are similar to those found by Hooftman and de Raat (1982) in *Umbra pygmaea* on exposure to ethyl methanesulphonate; by Rishi *et al.* (2001) on *C. punctatus* while using 2,4-D and butachlor and Sahoo and Bhuniya (2001) in *Heteropneustes fossilis* during exposure to carbaryl.

Hooftman and de Raat (1982) observed an increase in percentage of MN with increasing concentrations and durations of exposure to ethyl methanesulphonate in *Umbra pygmaea.* During present study significant increase in the number of MNC with increased concentration was found up to 96 hrs for 1 mg/l and up to 72 hrs for 2 mg/l. After 96 hrs at a concentration of 2 mg/l, a decline in number of MNC was observed (Table 2.1). The kidney cells were used during present work whereas Hooftman and de Raat (1982) used peripheral erythrocytes (PEC).

The effect of 2,4-D and butachlor in *C. punctatus* was made by Rishi *et al.* (2001). They reported a regular rise in number of MN formation from 24 to 72 hrs with 2,4-D and up to 48 hrs with butachlor treatment. But at 96 hr exposure the MN declined in both the exposures. In present findings, rise in number of MNC was observed from 24 to 96 hrs with 1 mg/l concentration and up to 72 hrs with 2 mg/l concentration but after 96 hr exposure the MN decreased for 2 mg/l concentration (Table 2.1).

Number of MN was increased with increased durations and concentration by Sahoo and Bhuniya (2001) in *Heteropneustes fossilis* on exposure to carbaryl pesticide. The present results are also in agreement with them except that after 96 hr, the number of MNC declined in the exposure with 2 mg/l.

Gradual increase in MN and nuclear anomalies was also reported by Sarangi *et al.* (2001) in *C. punctatus* during treatment with increased concentration of malathion for different durations (0, 5, 10, 15, 20, 25 days). The increase in MN was up to 20 days only and on 25th day a decline in number of MN was observed.

Watanabe *et al.* (1979) studied mutagenic effects of cadmium on mammalian oocyte chromosomes of female golden hamster and observed that the oocytes of this species were especially sensitive to cadmium and showed several instances of hyperploidy and diploidy in metaphase stages I and II.

Seoane and Oulout (1999) studied the aneugenic and clastogenic effect of $CdSO_4$ in Chinese hamster ovary cells and observed a high frequency of cells with chromatid bridges, lagging chromosomes and lagging fragments.

Pragya Khanna *et al.* (2005) studied the effect of $CdCl_2$ on Polytene chromosomes of *Chironomous plumosus* Form B for two successive generations. The results suggested that exposure to Cd enhances the activity of Chironomid genome and the resulting changes in the polytene chromosomes provide a subtle indicator of environmental stress.

Liu *et al.* (2006) showed that endoplasmic reticulum stress response plays a role in protection against cadmium-cytotoxicity in the LLC-PK1 renal epithelial cells.

Thus it can be concluded that the significant increase in the frequency of MN with increased duration and sub lethal concentrations shows that Cd is highly genotoxic. Therefore this heavy metal must be checked and prevented from entering into different water bodies and ultimately into the living system.

References

Alink, G.M., Frederix-Wolters, E.M.H., Vander Gaag, M.H., Van der Kerkhoff, J.E. and Poels, C.L.M., 1980. Induction of sister chromatid exchanges in fish exposed to Rhine water. *Mut. Res.*, 78: 369–374.

Al-Sabti, K., 1986. Comparative micronucleated erythrocyte cell induction in three cyprinids by five carcinogenic/mutagenic chemicals. *Cytobios*, 47: 147–154.

Al-Sabti, K., 1991. *Handbook of Genotoxic Effects and Fish Chromosomes.* Jozef Stefan Institute, Jamova.

Al-Sabti, K. and Metcalfe, C.D., 1995. Fish micronuclei for assessing genotoxicity in water. *Mut. Res.*, 343: 121–135.

Belies, R.P., 1975. *Toxicology*. Macmillan, New York.

Das, R.K. and Nanda, N.K., 1986. Induction of micronuclei in peripheral erythrocytes of fish *Heteropneustes fossilis* by mitomycin-C and paper mill effluent. *Mut. Res.*, 175: 67–71.

Hooftman, R.N., 1981. The induction of chromosomal observations in *Notobranchus rachowi* (Pisces : Cyprinodontidae) after treated with ethyl methanesulphonate. *Mut. Res.*, 91: 347–352.

Hooftman, R.N. and de Raat, W.K., 1982. Induction of nuclear anomalies (micronuclei) in the peripheral blood erythrocytes of eastern mudminnow *Umbra pygmaea* by ethyl methanesulphonate. *Mut. Res.*, 104: 147–152.

Khanna, Pragya, Sharma, O.P. and Tripathi, N.K., 2005. Genotoxic effect of cadmium chloride on Polytene chromosomes of *Chironomous plumosus* Form B (Chironomidae : Diptera). *J. Cytol. Genet.*, 6(NS): 47–55.

Liu, F., Inangeda, K. Nishiati, G. and Matsuoka, M., 2006. Cadmium induces the expression of Grp 78, an endoplasmic reticulum molecular chaperone, in LLC–PK1 renal epithelial cells. *Environ Health Perspect.*, 114(6): 859–864.

Matsumoto, S.T., Mantovani, M.S., Mallaguti, M.I. and Marin-Morales, M.A., 2003. Investigation of the genotoxic potential of the waters of a river receiving tannery effluents by means of the *in vitro* comet assay. *Cytologia*, 68(4): 395–401.

Rishi, K.K. and Jain, M., 1998. Effect of toxicity of cadmium on scale morphology in *Cyprinus carpio* (Cyprinidae). *Bull. Environ. Contam. Toxicol.*, 60: 323–328.

Rishi, K.K., Rishi, S., Grewal, S. and Gill, P.K., 2001. Genotoxicity of 2,4-D and butachlor on the chromosomes of fish *Channa punctatus*. *Persp. Cytol. and Genet.*, 10: 781–790.

Sahoo, S.N. and Bhuniya, 2001. Evaluation of genotoxic potential of carbaryl sevin R (1-napthylmethylcarbonate) in *H. fossilis in vivo* test systems. *Persp. in Cytol,. and Genet.*, 10: 377–381.

Samiullah, Y., 1990. *Biological Monitoring of Environmental Contaminants: Animals*. MARC report number 37, Global Environmental Monitoring Programme.

Sarangi, P.K., Porricha, S.K., Patnaik, Rajlakshmi and Parshad, K., 2001. Genotoxicity of malathion in *C. punctatus* cultured *in vivo. Persp. in Cytol. and Genet.*, 10: 835–844.

Seoane, A.I. and Dulout, F.N., 1999. Contribution to the validation of the anaphase-telophase test: Aneugenic and clastogenic effects of cadmium sulphate, potassium dichromate and nickel chloride in Chinese hamster ovary cells. *Genet. Mol. Biol.*, 22(4): 359–368.

Sokolova, I.M., 2004. Cadmium effects on mitochondrial function are enhanced by elevated temperature in a marine poikilotherm, *Crassostrea Virginia* Gmelin (Bivalvia : Ostreidae). *The J. Experimental Biol.*, 207: 2639–2648.

Watanabe, T., Shimada, T. and Endo, A., 1979. Mutagenic effects of cadmium on mammalian oocyte chromosomes. *Mut. Res.*, 67(4): 349–356.

Zou, E. and Bu, S., 1994. Acute toxicity of Cu, Cd and Zn to the water flea, *Moina irrasa* (Cladocera). *Bull. Environ. Contam. Toxicol.*, 52: 742–748.

Chapter 3

Diel Vertical Migration of Zooplankton in Andaman Sea

I.K. Pai[1], Sameer Terdalkar[2], M.L. Pereira[2] and F.M. Morgado[2]

[1]Department of Zoology, Goa University, Goa, India

[2]Department of Biology, Aveiro University, Aveiro, Portugal

ABSTRACT

Several groups of zooplankton exhibit Diel Vertical Migration (DVM), during which, hours of darkness are spent near the surface of the water and the day light hours at deeper depths. Though many earlier workers (Gunther, 1953; Ussings, 1938; Raymond, 1939; Marshell, 1960; Koslow, 1979 (cf. Raymont, 1983); Castel and Courties, 1982; Paffenhoper, 1983; Piatskowski, 1985; Mackas and Anderson, 1986; Ryan *et al.*, 1986; Kimmerer and McKinnen, 1987; Okemwa, 1989; Valdes *et al.*, 1990; Chakley, 1992; Meckas, 1992; Lafontaine, 1994; Haywood, 1996) have worked on the DVM all several regions of the world, the phenomenon is said to poorly understood (Raymont, 1983).

Andaman Sea is partially isolated part of north eastern Indian Ocean, which lies enclosed between the coasts of Myanmar. Thailand and Malaysia on east and chain of Andaman and Nicobar islands in Sumatra in the west, occupies 6.6 × 10^3 km^3 with an average depth of 1096M, is connected to Bay of Bengal by numerous channels. It is oligotrophic in nature with low primary and secondary productivity. It is also known to be rich in marine wealth and hence, attracted many researchers and several cruises in this area, dating back from 1869 lead by Francis Day. Inspite of this, due to poor sampling, as rightly pointed out by McAllister *et al.* (1994), not much is known either about the diversity or DVM of zooplankton in this area.

Hence, to fill this lacunae of information, an attempt has been made to study the DVM at Andaman Sea. ORV Sagar Kenya was made to station at a particular place and 12 samplings of 3 replicates each at an interval of two hours for entire diel cycle was conducted, by using Bongo net. The samples were collected from 30–OM depth. They were subjected for analysis for biomass, wet weight, species composition, species diversity, patchiness, evenness fluctuation and later the data obtained thus was subjected for cluster analysis (Similarity matrix method) and dendrograms were prepared to compare the similarity index with the time of sampling. The results obtained are discussed in the light of available literature.

Keywords: Andaman sea, Zooplankton, Species diversity, Diel vertical migration (DVM), Biomass, Wet weight, Patchiness, Evenness.

Introduction

Migration in zooplankton may be studied under two broad headings, on time scale-seasonal/annual migration, as studied by Turner (1982), Villatte (1991) related to life history patterns and diurnal pattern in which hours of darkness are spent near to the surface and day light hours at deeper depths as studied by Castel and Courties, 1982; Paffenhoper, 1983; Pialskowski, 1985; Mackas and Anderson, 1986; Ryan *et al.*, 1986; Kimmerer and MoKinnen, 1987; Okemwa, 1992; Valdes *et al.*, 1990; Checkley, 1992; Mackas, 1992; Lafontaine, 1994; Heywood, 1996), which is said to be beneficial for the organism (Ohman, 1990) and helps in their population budget and diurnal deficiency (Desstasio, 1993).

During DVM, members of zooplankton move as much as 400M an average small body sized species and over 600M in larger ones. They move rapidly at a speed of 12–200M per hour, which is about 45,000 times of body length per hour which would equal to a speed of 81 Km/h for a human of 1.8M height (Durbaum and Kunnemenn, 1999). The DVM may some times be undertaken twice a day in a diel cycle. During day time, they live in deeper depths of water, but during dark they ascend to surface. They then disperse through the water column in the middle of the night, phenomenon termed as 'midnight sinking' as explained by Russel, 1937 (cf; Raymont, 1983). They rise towards the surface again just before dawn, which is termed as 'dawn rise'. Some species may also exhibit reverse migration as reported by Parsons and Takahashi (1979). Further, the plankton can also be classified as drifting and residual one based on their behaviour (Kaartvedt, 1993).

Several workers, working in different parts of the world, have assigned various reasons for phenomenon of DVM in zooplankton (Zaret and Suffern, 1976; Enright, 1977; Stich and Lampret, 1981; Hauris, 1988; Bollens and Frost, 1989a,b,c; Magnesen, 1989; Neuman, 1989; Hansson *et al.*, 1990; Ohman, 1990; Jerling and Wooldridge, 1992; De Stasio, 1993).

The Andaman Sea, which is located south eastern side of Bay of Bengal, which is partially isolated portion of south eastern Indian Ocean, lying enclosed between the coasts of Myanmar, Thailand, Malaysia on the east and chain of Andaman and Nicobar Islands and Sumatra in the west, and occupies 6.02×10^3 km^2 with a volume of $6.6. \times 10^5$ km^3 with an average depth of 1096M. Unlike temperate seas, where spring bloom with regards to biomass is recorded (Lignell *et al.*, 1993), Andaman sea shows some what constant biomass all through the year. Further, due to its oligotrophic nature, it has low primary and secondary productivity (Anonymous, 1981) and hence high diversity of life. In this region, apart from a few stray works (Bhattathiri and Devassi, 1981; Goswami, 1981; Madhupratap, 1981; Vijayalaxmi, 1981) not much work has been carried out on zooplanktology in general and DVM in particular. Hence, an attempt has been made to study DVM of zooplankton in Andaman Sea.

Materials and Methods

During the Department of Ocean Development (Government of India) and National Institute of Oceanography, Goa organised multidisciplinary cruise SK-118 on ORV Sagar Kanya, which had predetermined area of operation as Andaman sea, the ship was stationed at 10°30'237"N latitude and 93°15'257"E longitude, where the depth was about 2300M, and the experiments were conducted.

The zooplankton samples required for the present studies were collected for a period of 24 hrs at an interval of 2 hrs, from the upper most euphotic zone, *i.e.*, from 30M to surface by Bongo net (Dia 0.6M, length 2.5M and mesh width 300 μm). A pre-calibrated flow meter (T.S. flow meter model no. 4512) was attached to calculate the amount of water filtered through the mouth of the net.

The zooplankton biomass was determined by displacement volume method and collected samples were preserved in 4 per cent formaldehyde. The wet weight was determined by following the method mentioned by Omori and Ikeda (1984). Later the samples were brought to the laboratory and analysed for abundance of major zooplanktonic groups and identified up to species level by using available literature (Kasturirangan, 1963; Mori, 1964; Daniel, 1985; Zheng Zhong, 1989 and Santhanam and Srinivasan, 1994). The diversity was calculated by using the indices given by Margalef (1968) for species of zooplankton encountered in the samples. All the zooplankton groups as well as species under study were analysed for their presence/absence in every sample of day/night individually and separately.

The 12 sample sets with three replicates were subjected for linkage clustering within the similarity matrix, which is considered as a convenient method for illustrating relationships. The method used was simple linkage method as described by Omori and Ikeda (1984).

Results

The samples collected during the entire diel cycle of 24 hrs with an interval of 2 hrs showed the biomass fluctuating from 0.01 m/m^{-1} to 0.05 m/m^{-1}, with an average of 0.018 m/m^{-1}. The night samples exhibited higher biomass ranging from 0.02 to 0.05 m/m^{-1} with an average of 0.028 m/m^{-1}. Thus showing higher quantity and concentration of zooplankton at 0–30 M depth at night (Figure 3.1).

Similarly wet weight varied from 0.4 to 2.2 mgm^{-3} (Figure 3.2), with an average of 0.65 mgm^{-3} for day samples and from 1.0 to 2.2 mgm^{-3} (average 1.55 mgm^{-3}) for night samples (Figure 3.2), thus recording considerably higher wet weight for night samples.

The zooplankton groups encountered in the samples were copepods, chaetognaths, emphipods, crustaceans, siphonophores, pelagic tunicates, fish eggs/larvae etc.

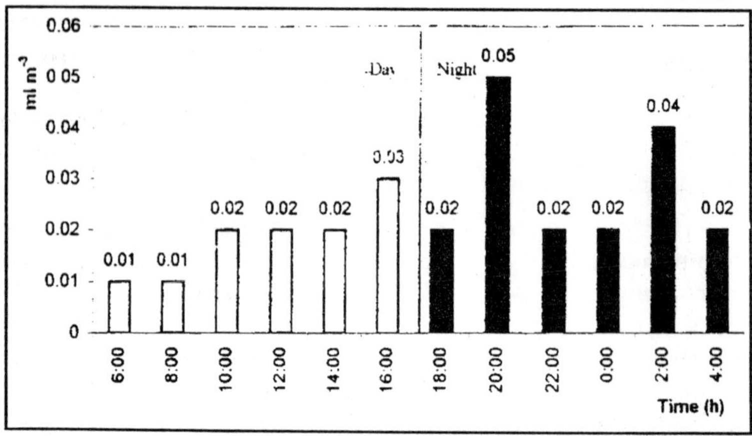

Figure 3.1: Biomass

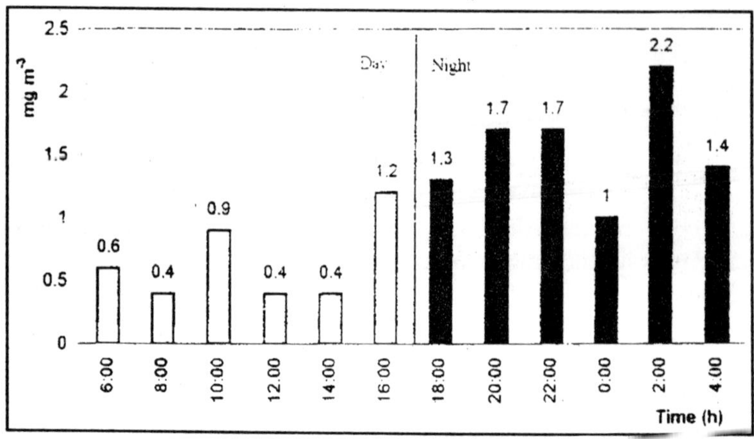

Figure 3.2: Wet Weight

Though more than 65 species of copepods such as *Candacia pachydactyla, Eucheata concinna, Sapphiria negromaculata, Copilia mirabilis, Rhinocalanus nasutus, Corycaeus affinis, Oncaea venusta, Oncaea media, Pontellopsis tenuicaudata, Pontellopsis sacurifer, Schmackria popleisa, Subeucalanus longiceps, Unidula vulgaris, Acartia* spp., *Clausocalanus breviceps, Sinocalanus* spp., *Acartia tonsi, Corycaues speciosus, Thysanopoda tricuspidata, Stylocherion carinatum, Chaelophyses appendiculata* etc., were encountered in the studies, only first five species mentioned above have been considered for present detailed DVM studies.

Figure 3.3 demonstrates the percentage of copepods in entire day and night collections. It can be seen that during day time, their percentage ranged from 54.0 per cent to 82.0 per cent (Average 70.66 per cent), of total catches, while at night it was ranging from 42.0 per cent to 65.0 per cent (Average 55.5 per cent) indicating, comparatively low percentage of their existence at upper layer of water at night.

Figure 3.4 explains the percentage of *Candacia pachydactyla* in the entire diel cycle. It clearly exhibits two peaks of its existence (one

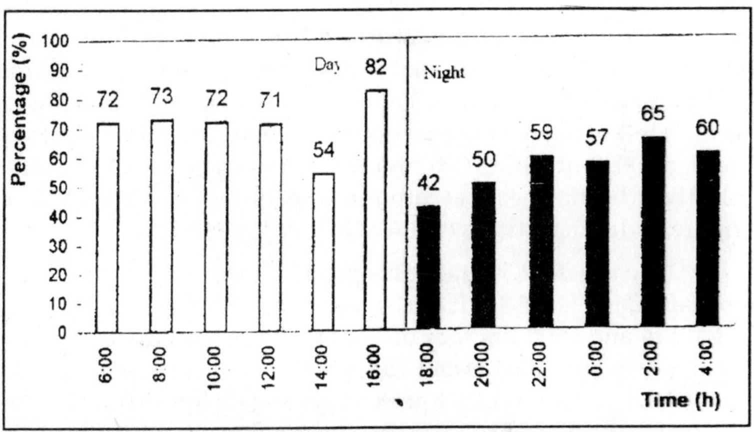

Figure 3.3: Percentage of Copepods in Collection

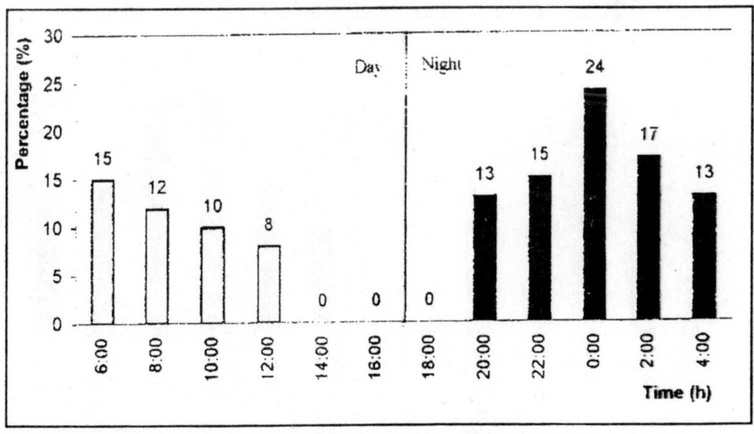

Figure 3.4: Percentage of *Candacia paechydactyla* among Copepods

at 0600 hrs and other at 0000 hrs), exhibiting its diel vertical migration pattern. Their existence, in the catches ranged from 0 per cent (at 1400 hrs and 1600 hrs) to 24 per cent at 0000 hrs. Further, at an average, they were found in relatively abundant in night collection than daytime collection.

Eucheata concinna, though exhibits its presence in upper water column (0-30M) an through the diel cycle, it ranges at an average of 9.33 per cent at day to 13.33 per cent at night, with their day peak at 1400 hrs and night peak of their existence at 2200 hrs, thus exhibiting DVM (Figure 3.5).

Figure 3.6 shows the existence of *Sapphirina negromacculata* in 0-30M water column. During the day it ranges from 0 per cent (at 1600 hrs) to 6.0 per cent (at 1400 hrs) with an average of 3.66 per cent. While at night it ranges from 0 per cent (at 1800 hrs) to 12.0 per cent (at 0000 hrs and 0200 hrs) with an average of 6.0 per cent, clearly indicating peaks at corpuscular period of the diel cycle. The figure also demonstrates the DVM in *S. negromacculata.*

Diel vertical migration of copepod *Copilia mirabilis* has been explained in Figure 3.7. It can be seen that at 1400 hrs and between 0400 hrs and 0600 hrs they are not found in the surface waters. While, between 0800 hrs and 1200 hrs as well between 1600 hrs and 2000 hrs they are found at a percentage ranging from 9.0 per cent to 12.0 per cent. Further it can also be seen that whenever they are there, their existence is almost similar.

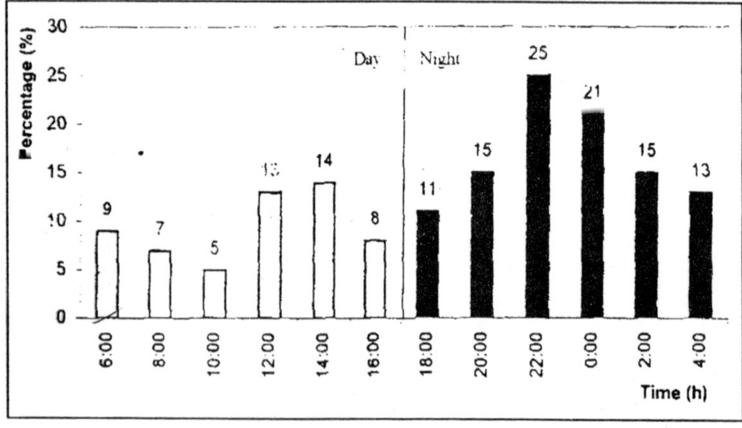

Figure 3.5: Percentage of *Eucheata concinna* among Copepods

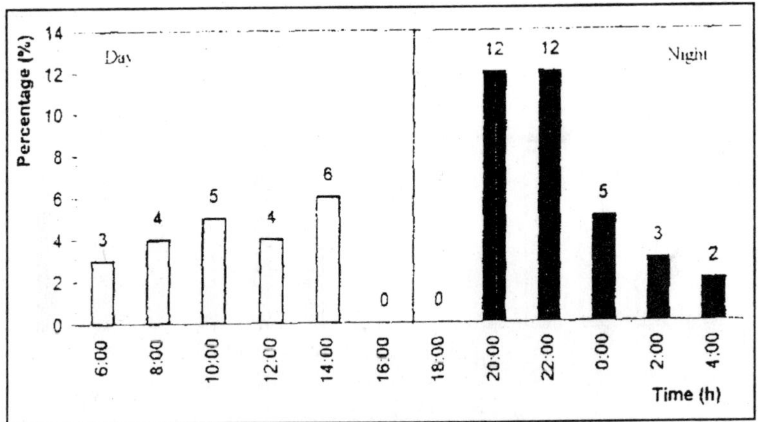

Figure 3.6: Percentage of *Sapphirina negromcullata* among Copepods

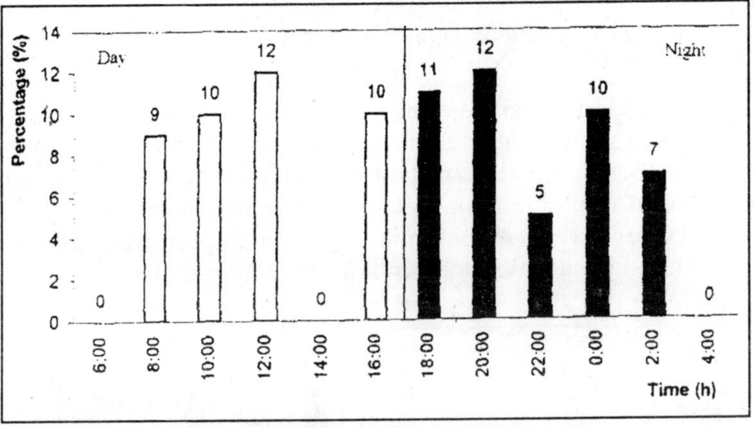

Figure 3.7: Percentage of *Copilia mirabilis* among Copepods

Rhinocalanus nasutus, though exhibits its presence in surface waters at all the time, higher percentage of 22.0 per cent and 20.0 per cent was seen at 1400 hrs and 0200 hrs respectively, while the minimum of 5.0 per cent and 7.0 per cent between 1600 hrs and 1800 hrs and 11.0 per cent and 12.0 per cent between 0400 hrs and 0800 hrs, clearly exhibiting two diel cycles (Figure 3.8).

Though individuals belonging to 10 species of amphipods were collected during the studies, *Brachyscelus crusculum, Lestrigonus*

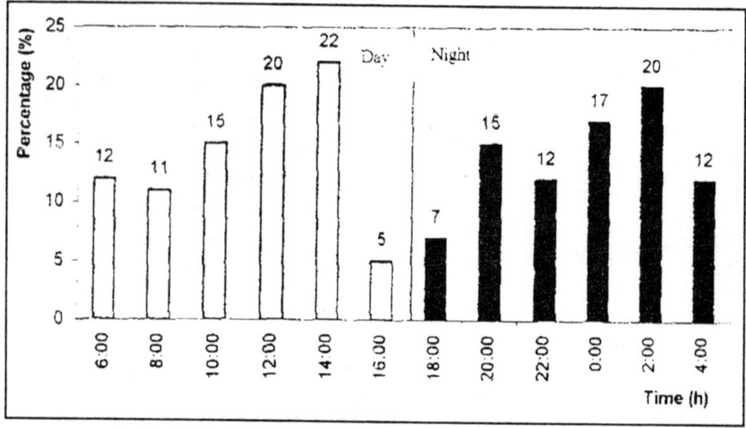

Figure 3.8: Percentage of *Rhinocalanus nasutus* among Copepods

schizogeneios, *Cranocephalus* spp. and *Rhabdosome* spp. were prominent ones.

Amphipods seem to come to the surface waters only at night. They also show a clear "mid-night sinking". It can be seen from Figure 3.9, amphipods start showing their presence in 0-30M water column at around 1600 hrs with peak immediately after sun set, *i.e.*, at 1800 hrs and later show reduction in their population and by 0200 hrs show a "mid-night sinking" to re-appear at 0400 hrs.

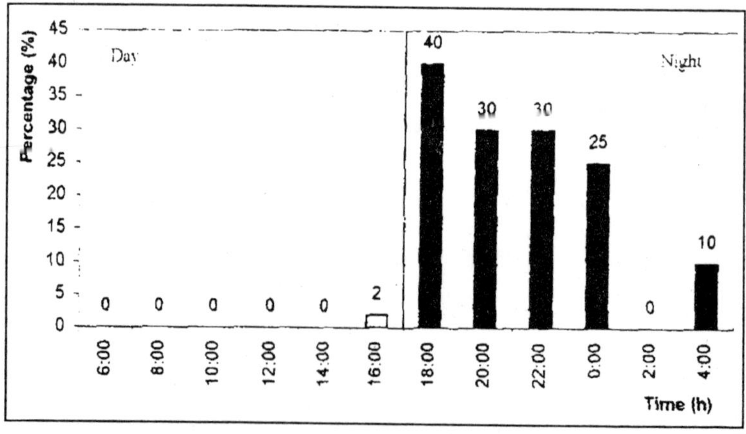

Figure 3.9: Percentage of Amphipods in Collection

Figure 3.10 indicates the pattern of DVM of amphipod *Brachyscelus crusculum,* which shows that the organism comes to the surface water only between 1800 hrs and 0000 hrs and they complete their feeding and other activities in about 6 hrs and they move down 10 spend rest of the part of their diel cycle.

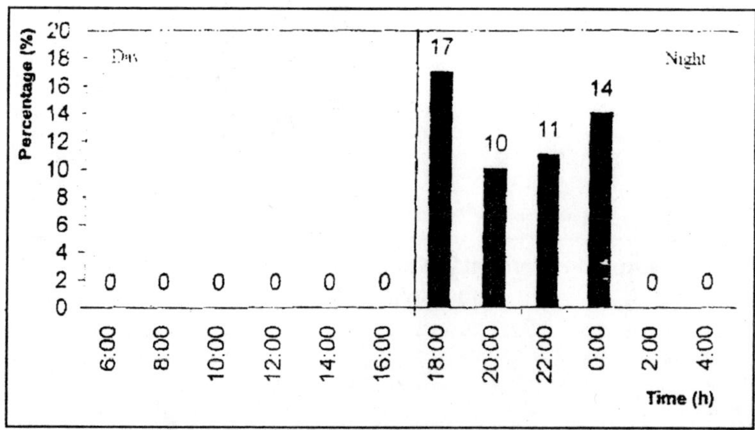

Figure 3.10: Percentage of *Brachyscelus crusculum* among Amphipods

Similarly, *Lestrigonus schizogeneios* shows its presence in 0-30M water column from 1800 hrs to 0000 hrs and after completing their feeding of phytoplankton they sink down to spend other part of their diel cycle (Figure 3.11), like other amphipod *Brachyscelus crusculum.*

Amphipod *Cranocephalus* spp. also exhibits its presence in 0-30M water column between 1800 hrs to 0000 hrs. After peak existence (26.0 per cent among amphipods) they sink down to spend other time of their diel cycle (Figure 3.12).

Rhabdosoma spp. also exhibit similar pattern of diel cycle as that of *Cranocephalus* spp., *Lestrigonus schizogeneios* and *Brachysceles crusculum* by spending 1800 hrs to 0000 hrs at surface waters with maximum density of 27 per cent among amphipods at 2200 hrs and spend rest of the time of their diel cycle at deeper water column (Figure 3.13), thus indicating that all four species under study under take a single DVM cycle a day.

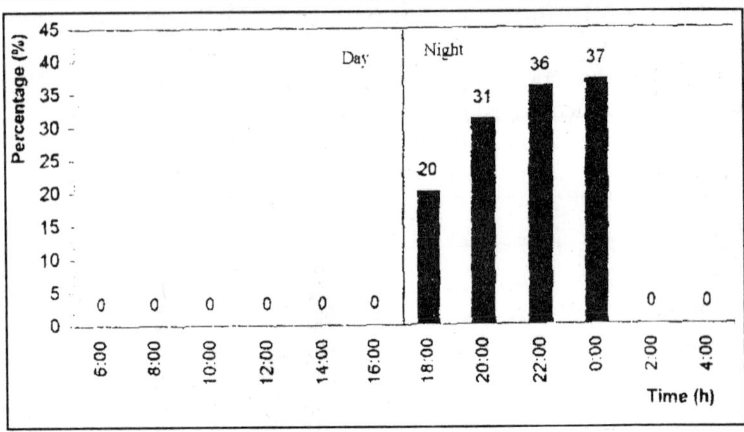

Figure 3.11: Percentage of *Lestrigonus schizogeneios* among Amphipods

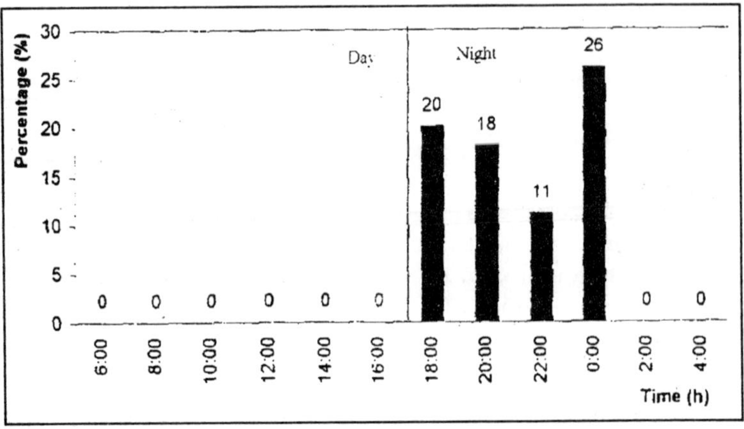

Figure 3.12: Percentage of *Cranocephalus* spp. among Amphipods

Cheatognaths in general, as they have larger body size and prominent eye, seems from Figure 3.14, that they spend more day time at surface waters. But they too show a peak existence of 30 per cent at 0800 hrs and 0200 hrs, white a minimum of 0 per cent was seen at 1800 hrs indicating their cyclic movements.

Pterosagitta draco, a chaetognath with larger body size and large eyes seems to prefer grazing at upper layers during 0600 hrs to 1400 hrs, when there are less plankton grazers at that time. Figure 3.15

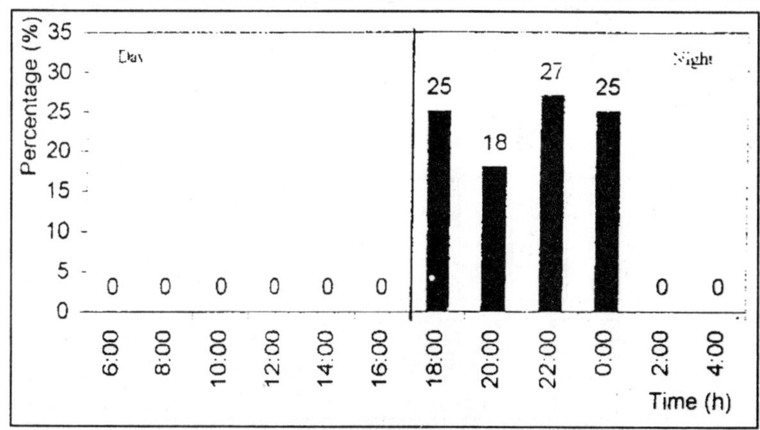

Figure 3.13: Percentage of *Rhabdosoma* spp. among Amphipods

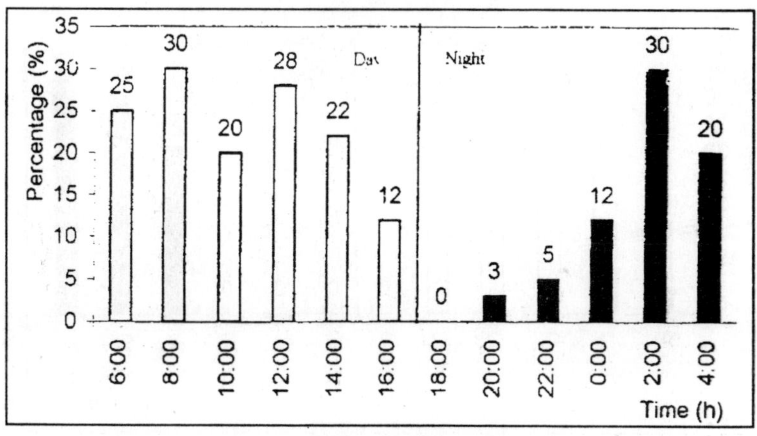

Figure 3.14: Percentage of Chaetognaths in Collection

demonstrates *P. draco*'s presence at 23.0 per cent to 43.0 per cent from 0600 hrs to 1400 hrs and later they move down to lower water column to spend rest of the time of their diel cycle. It looks as if they follow one DVM cycle a day.

Sagitta robusta, one more example for chaetognath, shows its presence during day time as well at night time (Figure 3.16), but at relatively a lower percentage (15.0 per cent to 30 per cent) among chaetognaths at day compared to 31.0 to 44.0 per cent at night

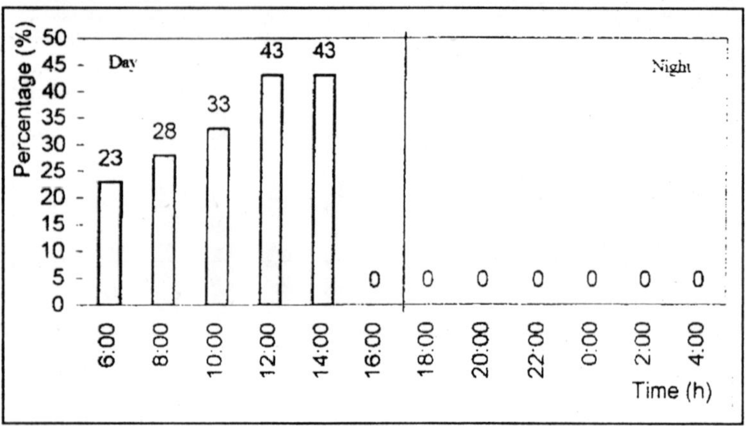

Figure 3.15: Percentage of *Pterosagitta draco* among Chaetognaths

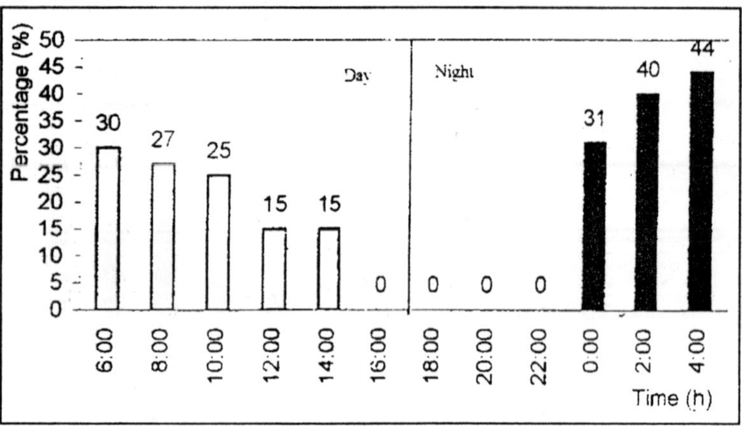

Figure 3.16: Percentage of *Sagitta robusta* among Chaetognaths

whenever they are present. Apart from the above, it can also be seen from Figure 3.16, that they move to lower waters from 1600 hrs to 2200 hrs, to demonstrate that they too undertake diel vertical migration, though once a day. Further, it seems, both *P. draco* and *Sagitta* spp, to avoid competition among themselves, size of the populations present during daytime is inversely proportional to each other.

Sagitta bipuncta, also falls in line with *S. robusta* with regard to its diel vertical movements. It can be seen from Figure 3.17, the peak existence of this organism among chaetognaths is 50.0 per cent at 0400 hrs while from 1400–2200 hrs, they do not show their existence in 0-30M water column, indicating that they too have one DVM cycle per day. As there is no existence of *P. draco* at night, probably there is sufficient food for both *S. robusta* and *S. bipunctata*, if they are feeding on same kind of food, or else, they must be having different feeding habits, to avoid competition for same source.

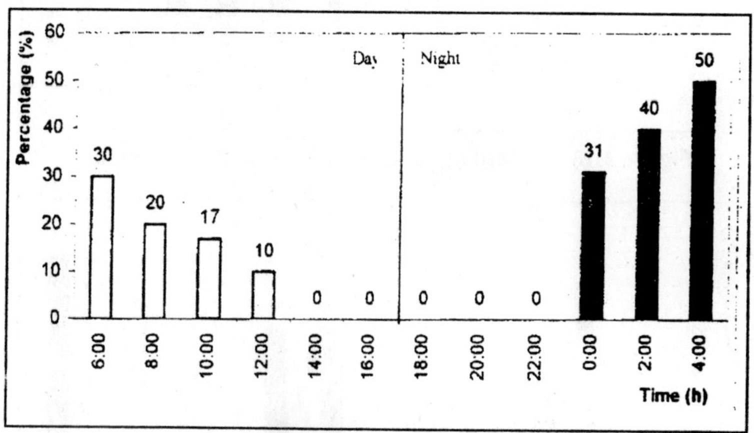

Figure 3.17: Percentage of *Sagitta bipunctata* among Chaetognaths

Figure 3.18 exhibits the percentage of adult crustaceans among the total collection. It ranged from 1.0 per cent to 10.0 per cent between 1600 hrs and 0000 hrs and later they sink down. Thus exhibiting a diel migratory cycle.

Siphonophores, clearly show two cycles of their vertical migration in a day with zero presence between 1200 hrs and 1600 hrs and between 2200 and 0200 hrs, with peak presence of 10 per cent at 1800 hrs and 2000 hrs and 14.0 per cent at 0400 hrs (Figure 3.19). It can also be seen that they prefer night (presence 10.0 per cent to 14.0 per cent) than day (presence 3.0 per cent to 5.0 per cent) time to come to surface of the water. Further, by going through Figure 3.19, one can see that they appear at corpuscular period of the day, when most of the other predator species like amphipods and chaetognaths

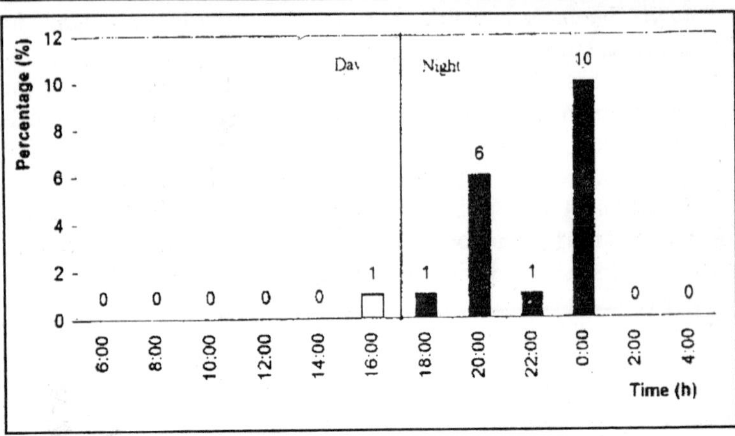

Figure 3.18: Percentage of Crustaceans in the Collection

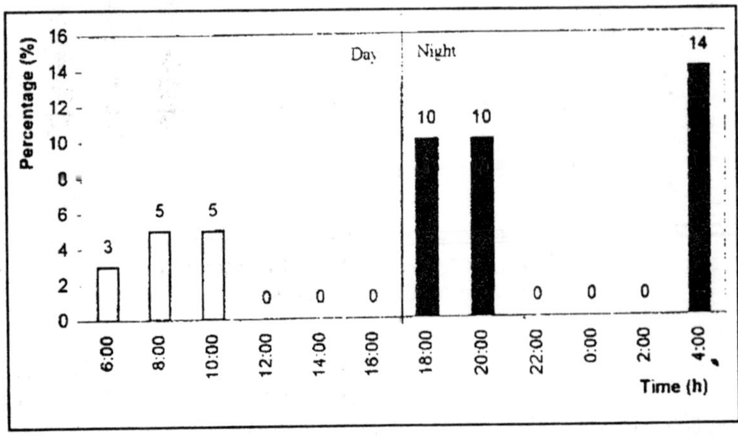

Figure 3.19: Percentage of Siphonophores in the Collection

are absent or at a low density, though the siphonophores are also predators in general, but with smaller body size.

Tunicates spend less time of about 6 hrs in 0.30M water column that too after sunset and finish their activity quickly and sink down (Figure 3.20).

Figure 3.21, explains the diel fluctuations of species diversity index for zooplankton in Andaman Sea. It is interesting to see that, though the zooplankton exhibit vertical migration, and the range of species diversity varies from 6.7 at 0200 hrs to 11.3 at 1600 hrs, with

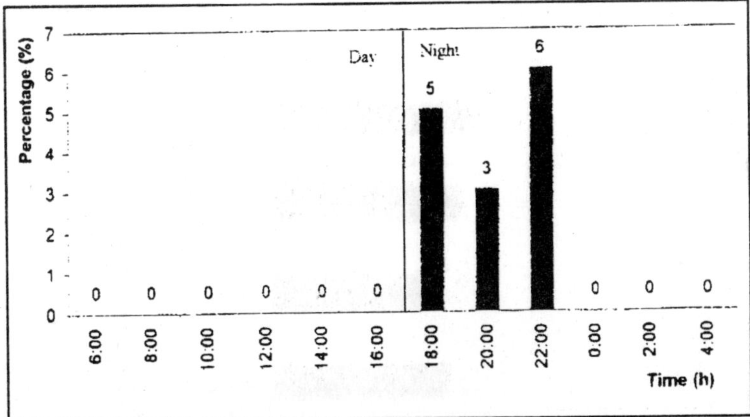

Figure 3.20: Percentage of Tunicates in the Collection

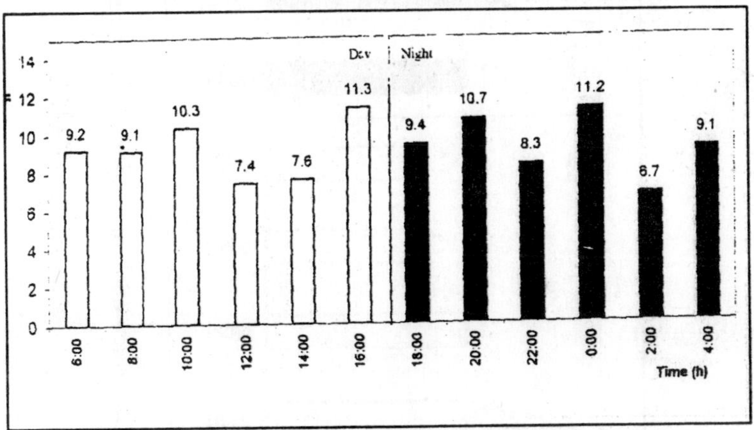

Figure 3.21: Diel Fluctuation of Species Diversity Index of Zooplankton in Andaman Sea

an average of 9.15 and 9.23 diversity index for day and night respectively, which is all most similar.

Diel fluctuations with regard 10 evenness of zooplankton at Andaman Sea is represented in Figure 3.22. It can be seen that evenness varies during day as well at night. The maximum evenness was 2.0 at 0800 hrs for day and 2.2 at 2000 hrs for night and the least evenness for day time was 1.3 at 1400 hrs, white for night it was 1.5 at 0000 hrs, indicating the fluctuation of populations, thus providing support for vertical migration.

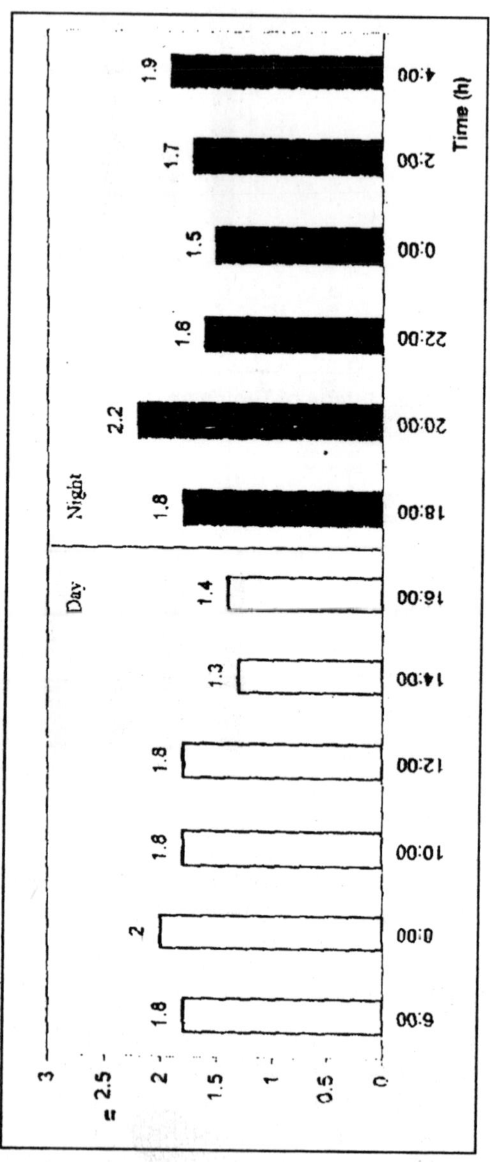

Figure 3.22: Diel Fluctuation of Evenness of Zooplankton in Andaman Sea

The result of cluster analyses and dendrogram comparing similarity matrix with time of sampling. It can be seen that the highest value of 0.91 for sample collected at 1200 hrs and 1000 hrs thus linking them together at 0.91 level. Second highest similarity was at 0.81 between samples collected at 0600 hrs and 0800 hrs thus linking them together. Third highest link was at 0.78 between the sample collected at 0800 hrs and 1000 hrs, thus forming a group. The next groups was at 0.66 level by the samples collected at 2000 hrs and 2200 hrs and finally the least similarity value was 0.25 recorded for the samples collected at 1600 hrs and 1400 hrs.

Discussion

Biomass and wet weight analyses are two well known yard sticks to analyse the density of living organisms, generally in marine environment. Figures 3.1 and 3.2 clearly shows that both biomass and wet weight is higher for samples collected during night time, than the day samples, indicating higher density of organisms at night. Lesser biomass and wet weight of the samples during the day time indicates that the organisms descend down to lower depths for several reasons as mentioned by earlier workers (Zaret and Suffern, 1976; Enright, 1977; Stich and Lampret, 1981; Hauris, 1983; Bollens and Frost, 1989a,b,c; Magnesen, 1989; Neuman, 1989; Hansson *et al.*, 1990; Ohman, 1990; Jerling and Wooldridge, 1992; De Stasio, 1993), one of the major reason is said to be avoidance of predators. The second is said to be passively to facilitate the phytoplankton to grow during day time thus to ensure continuous food for zooplankton.

In animal kingdom, members of different groups exhibit distinct behaviours, which are specific to a particular group, though among the members of the slame group one can find variation with regard to that behaviour. Diel Vertical Migration (DVM) though known to be prominent and well studied in zooplankton by earlier workers like Gunther, 1935; Ussings, 1938; Raymond, 1939; Marshall, 1960; Koslow, 1979 (Cf: Raymont, 1983) is poorly understood. Several workers have assigned various reasons for migratory and non-migratory behaviour of zooplankton (Vinogradov, 1970; Raymont, 1983; Chae, 1995). One of the major cues for migration has been assigned to light intensity (Neuman, 1989; Jerling and Wooldridge, 1992), and have suggested while changing light intensity is significant for migration as "optimum zone" of light intensities is not only the determining factor. Animal has to either adapt to changing light intensity during the day, which will trigger off other behavioural patterns, like copepod in general and *Euchaeta concinna* and *Rinocalanus nasuta* in particular which show their presence throughout the diel cycle, though in varied degree. But among copepods, one can also see the species like *Candacia polydactyla, Sapphirina negromaculata, Copilia mirabilis,* which can not adopt to surface waters all the time. Hence, they sink to lower depths between 1400 hrs and 1800 hrs, 1600 hrs–1800 hrs and around 1400 hrs respectively, indicating variation in adaptability among the species of same group. While some other species like *Schmackeria poplosia, Subeucalanus longiceps, Unadula vulgaris* etc., appeared at the surface waters only for a short period, that too during 2200 hrs to 0600 hrs. The reason for this may be that, the daytime distribution of sapphirinids are determined under water light conditions as proximal cue as hypothesised by Chae (1995), that the well developed eyes, the irridescence of the males and day time shoaling in sapphirinids are closely related and constitute presumed mate finding mechanisms, which may be unique in oceanic planktons. Further, as explained by Reymont (1983) marked pigmentation reduces the photodamage to copepods and make it easily visible to predators. This seems to be the case with copepods like *Subeucalanus longiceps, Unadula vulgaris,* which remain in lower depths, where the light intensity if low during day time and surface at night.

Regarding amphipods, from the present experiments, it is seen that they come to the surface water twice a day, once for long duration from around 1600 hrs to 0000 hrs and after a short "midnight sinking" reappear for a short period at around 0400 hrs and then they will have a long sinking from 0600 hrs to 1600 hrs may be to avoid predators at day time, or this may also be a natural phenomenon, where nature allows the phytoplankton to grow during the day time when sunlight is available, to ensure continuous food supply as suggested by Hauris (1988).

About Amphipods, all the species under study (Figures 3.10–3.13) showed their existence in 0.30M water column from 1800 hrs to 0000 hrs indicating their group behaviour. All the species under study showed simultaneous appearance for about 6 hrs and after completing their feeding and other activities at upper layer seems to descend to lower levels together, to re-appear in next cycle. As the size of the organism is relatively small, it seems to avoid the predators at day time, they do not come to surface layers. Further, as they show a short period of grazing, due to less consumption, to save energy, they may be sinking down to lower water column, where metabolic rate is low due to low temperature thus needing less energy requirements as suggested by De Stasio (1993). Further, appearance and disappearance of amphipoda as a group may be a strategy to make way to other groups to come to surface layers for feeding. It seems like a natural adjustment and adaptability phenomenon to reduce competition for same resource as suggested by Rothaupt (1990).

It is interesting to see from Figure 3.14, that though chaetognaths, a relatively larger animals with prominent eyes can afford to stay at upper layer of waters during day as well at night time, but indicates that they too undertake DVM, which is evident from total absence of chaetognaths at 1800 hrs and re-appear at a later stage to show their maximum percentage (30 per cent) at 0200 hrs. They also show the fluctuation with regards to their percentage of existence, though they exhibit their existence all through the daytime.

It can be seen from Figures 3.9 and 3.14 that, when amphipod population decreases from 40.0 per cent to 0 per cent from 1800 hrs to 0200 hrs, the chaetognath population increases from 0 per cent to 30 per cent, indicating that they avoid competition in grazing the same ground at the same time.

However, it can also be seen that among chaetognaths too seemingly to avoid inter and intra-specific competition, they distribute themselves during day and night. For *e.g. Pterosagitta draco* surfaces for grazing at day time and do not show its existence at night (Figure 3.15), while *Sagitta robusta* shows its presence day as well at night. With a zero presence from 1600 hrs to 2200 hrs (Figures 3.16) the *Sagitta bipunctata,* shows its presence from 0000 hrs (31.0 per cent) and peaks to 50.0 per cent at 0400 hrs and then reduces its presence steadily from 0600 hrs to reach 0 per cent at 1400 hrs. Thus not only reducing the competition among the members of the same groups but also giving enough time for phytoplankton to grow during the day time.

When copepods, amphipods, chaetognaths, tunicates, siphonophores are not present at their maximum concentration (Figures 3.3, 3.9, 3.14, 3.19, 3.20) adult crustaceans are found to be present at their maximum at the surface layer (at around 0000 hrs). Again supporting the evidence for avoidance of competition for same source of food as reported by Rothaupt (1990). As the crustaceans are generally smaller in size and exhibit a prolonged larval stages, it seems to avoid of being predated, they start surfacing from 1600 hrs, when the light intensity reduces and spends time upto 0000 hrs and then sink to lower layers of water.

While the siphonophores, by showing their presence at 0-30M water column at 1800 to 2000 hrs and 0400 to 1000 hrs clearly exhibit two diel cycles of their vertical migration (Figure 3.19). They exhibit midnight sinking too (Figure 3.19), it can also seen from the above, that they show their existence at day and night, their density demonstrates that they prefer night time than day. From the collections it was also observed that day time surfacers have relatively larger body size and large eyes, indicating that the body size has a bearing on day/night migration as reported by Mackas (1992). Further, siphonophores appear when competition pressure from other fellow predators is less, which is evident from Figure 3.19. Apart from that, they appear strictly during corpuscular hours of the diel cycle, when most of the other larger predator species such as amphipods, chaetognaths are either absent of in their low population. This may he a natural phenomenon of avoiding competition among various groups of organism for same source of food.

From Figure 3.20, it can be concluded that tunicates have relatively small range of timing for feeding and other activities at surface waters. As they are not known as long migrants, show a quick appearance at 1800 hrs and sink down to certain level at 2000 hrs to aappear at 2200 hrs and sink down to re-appear at next cycle. The short period of spending time at surface layer seems to either they are quick feeders or feed on surface phytoplankton to a less extent and prefer to stay at lower water columns may be safety or for their probable capacity to feed at that level too.

It is Pielou (1966) explained the importance of study of species diversity and its patter while studying ecological succession and later Peel (1914) in his review, described role and method of measuring species diversity. Hence, based on the importance, when an attempt was made to analyse the zooplankton diversity in the study area, the results obtained has been represented in Figure 3.21.

It can be seen from the figure, though there is a species diversity fluctuation during day or night time, a some what constant number of species (9.15 during day and 9.23 during night) was recorded, there is a record of higher species diversity during corpuscular period. Similarly patchiness also shows, its higher value at corpuscular period. Though the biomass of zooplankton at night was comparatively higher (2.83 m/m^{-3}) than the day time (1.83 m/m^{-3}), which may be a strategy of nature to see that all planktonic groups get an opportunity to graze, through at different times of the diel cycle. At the same time, by keeping low grazing pressure at day time, provides sufficient time for phytoplankton to grow in order to provide a continuous food supply for these zooplankton, thus maintaining the food chain in an efficiently operating manner. Thus indicating that Andaman sea ecosystem sustains a large, mature community with stability, where in a number of organisms live in absolute harmony for the their as well as community success.

Similarity index, patchiness and evenness are important criteria to determine the pattern of distribution of any organism (Omori and Hamner, 1982; Washington, 1989; Davies *et al.*, 1991). Hence, the results obtained in the present experiments were subjected for cluster analyses and dendrogram were prepared from comparing similarity index. The results obtained are represented, which clearly demonstrates, that there is a distinct grouping with regard to cluster formation in terms of similarity matrix. For example sampling at 1200 hrs and 1000 hrs belong to same cluster, with similarity matrix around 0.91, between 0600 hrs and 0800 hrs (0.81).

From the present studies, it becomes clear that the biomass in Andaman Sea remains some what similar during entire diel cycle, though whose higher peak at 2000 hrs and 0200 hrs, mainly due to the presence of maximum number of zooplankton. While the wet weight of the collected sample is significantly high at night than the day.

Regarding zooplankton, copepod groups seems to occupy the surface waters at all the times with some what equal distribution, though certain species like *Candacia pachydactyla, Sapphirina negromacculata, Copilia mirabilis* exhibit a complete (2 cycles per day) diel vertical migration, *Rhinocalanus nasutus, Euchaeta concinna* show their cyclic migration once a day, thus leading to the fluctuation of population size at the surface layers.

Amphipods show a general tendency as a groups to be at the surface waters at 1800 hrs to 0000 hrs, as exemplified by *Brachyscelus crusculum, Lestrigonus schizogeneios, Cranocephalus* spp., *Rhabdosoma* spp. etc. There are other amphipod species which may surface slightly early (1600 hrs) or at a slightly later stage (at around 0400 hrs), but they do not surface between 0600 hrs to 1400 hrs.

Similarly chaetognaths show fluctuation with regard to their existence, they do surface at all the times of the diel cycle except at around 1800 hrs *i.e.*, immediately after sunset. There are chaetognaths which surface during day time (*e.g., Pterosagitta draco*) or at day and night (*e.g. Sagitta robusta* and *S. bipunctata*), with distinct sinking between 1600 hrs and 2200 hrs.

The Crustaceans have a short span of surfacing from 1600 hrs to 0000 hrs with peak at 0000 hrs, before sinking to re-appear at next cycle. While the siphonophores show clearly two cycles in a day by appearing between 0400 hrs and 1000 hrs and later between 1800 hrs and 2000 hrs. While the tunicates appear to surface waters only once a diel cycle from 1800 hrs to 2200 hrs.

The present studies indicate that at all given times, of a diel cycle an average of 9.16 species of zooplankton will be at surface compared to 9.23 species at night indicating no significant variation with regard to species diversity. While the density of zooplankton varies from 1.3 to 2.0 at day (Average 1.68) and between 1.5 to 2.2 at night (Average 1.78) indicating higher evenness of grazers at night.

The studies also leads one to conclude that at Andaman sea, the biotic community supports some what same number of species during day as well at night, though the biomass varies, indicating ecological and trophic maturity of the community, thus exhibiting equilibrium. One can also see the equilibrium between day and night time production. There also exists equilibrium for competition for same resource between and among the members of different groups of organism.

Another aspect revealed by the present studies is the formation of distinct clusters of similarity with the time of sampling of zooplankton. For example there is a day group (0400, 0600, 0800, 1000 and 1200 hrs) and an evening group (1600 and 1800 hrs), while the catches at 1400 hrs do not show clustering with any group. While the catches at 0200 hrs, based on their numerical biomass and wet weight, seems to be near to evening group.

Acknowledgement

It is pleasure to thank Mr. Sergio Leandro and Mr. Aldiro Pereira for their technical help. Some of us (IKP and ST) express our sincere gratitude to Department of Ocean Development (Government of India) and National Institute of Oceanography, Goa for providing cruise facilities.

References

Anonymous, 1981. The Andaman Sea. *Ind. J. Mar. Sci.*, 10: 209–210.

Bhattathiri, P.M.A. and Devassi, V.P., 1981. Primary productivity of the Andaman Sea. *Ind. J. Mar. Sci.*, 10(3): 243–247.

Bollens, S.M. and Frost, B.W., 1989a. Zooplanktivores fish and variable DVM of marine zooplanktonic copepods *Calanus pacificus*. *Limnol. Oceanogr.*, 34: 1072–1083.

Bollens, S.M. and Frost, B.W., 1989b. Predator induced diel vertical migration in a planktonic copepod. *J. Plankton Res.*, 11(5): 1047–1065.

Bollens, S.M. and Frost, B.W., 1989c. Diel vertical migration in zooplankton: Rapid individual response to predators. *J. Plankton Res.*, 13(6): 1359–1365.

Castel, J. and Courties, C., 1982. Composition and differential distribution of zooplankton in Arcachen Bay. *J. Plankton Res.*, 4(3): 413–433.

Chae, J., 1995. Vertical distribution and diel migration in the irridiscent copepods of the family Sapphirinidae unique example of reverse migration? *Mar. Ecol. Prng. Ser.*, 119(1–3): 111–124.

Checkley, D.M., Uya, S., Dagg., M.J., Mullian, M.M., Omori, M., Onbe, T. and Zhu, M.Y., 1992. Diel variation of zooplankton and its environment at neritic stations in the Inland sea of Japan and North West Gulf of Mexico. *J. Plankton Res.*, 44(1): 1–40.

Daniel, R., 1985. *Fauna of India: Coelenterata, Hydrozoa, Siphonophora.* Z.S.I., Kolkata.

Davies, C.S., Fliert, G.R., Wiebe, P.H. and Franks, P.J.S., 1991. Micropatchiness: Turbulance and recruitment of plankton. *J. Mar. Res.*, 49: 109–151.

De Slasio, Jr. B.T., 1993. Diel vertical and horizontal migration by zooplankton: Population budgets and the diurnal deficit. B. *Mar. Sci.,* 53(1): 44–64.

Durbaum and Kunnemenn, 1999. *Biology of Copepods: An Introduction.* www.unioldenberg.de/zoomorphology/biologyintro.html: 1–4.

Enright, J.T., 1917. Diurnal vertical migrational adaptive significance and timing, Part–I: Selective advantage: A metabolic method. Limon. Oceanogr., 22: 856–872.

Goswami, S.C., 1981. Copepod swarming in cambell bay (Andaman Sea). *Ind. J. Mar. Sci.,* 10(7): 274–275.

Hansson, S., Larsson, U. and Johansson, S., 1990 Selective predation by hering and mysids and zooplankton community structure in Baltic Sea coastal area. *J. Plankton Res.,* 12(5): 1059–1116.

Hauris, R.P., 1988. Interaction between diel vertical migratory behaviour and marine zooplankton and surface chlorophyll maximum. B. *Mar. Sci.,* 43(3): 663–674.

Heywood, K.J., 1996. Diel vertical migration of zooplankton in the north-east Atlantic. *J. Plankton Res.,* 18(2): 163–184.

Jerling, H.C. and Wooldridge, T.H., 1992. Lunar influence on distribution of a copepod in the water column of a shallow temperature estuary. *Mar. Biol.,* 112: 309–312.

Kaarivedt, B., 1993. Drifting and resident plankton. B. *Mar. Sci.,* 53(1): 154–159.

Kasturirangan, L.R., 1963. *A Key for the Identification of the more Common Planktonic Copepods of Indian Coastal Waters.* CSIR, New Delhi, India.

Kennish, M.J., 1989. *Practical Handbook of Marine Science.* CRC Press, Boston, p. 307–308.

Kimmerar, W.J. and McKinnon, A.D., 1981. Zooplankton in a marine bay. I. Horizontal distribution used to estimate shelf population growth rate. *Mar. Ecol. Prog. Ser.,* 41: 43–52.

Lafontaine, Y., 1994. Zooplankton biomass in the southern gulf of St. Lawrence: Spatial pattern and the influence of fresh water run-off. *Can. J. Fish Aquat. Sci.,* 51: 617–735.

Lignell, R., Heiskanen, A.S., Kuosa, H., Gundersen, K., Kuuppoleinikki, P., Paguniemi, R. and Uitto, A., 1993. Fate of phytoplankton spring bloom: Sedimentation and carbon flow in the Northern Baltic. *Mar. Ecol. Progr. Ser.*, 94: 239–252.

Mackas, D.L., 1992. Seasonal cycle of zooplankton off Southern British Columbia, 1979–1989. *Can. Fish. Aquac. Sci.* 49: 903–921.

Mackas, D.L. and Anderson, E.P., 1986. Small scale zooplankton community variability in Northern British Columbia Fjyord system. *Estuar. Coast. Shelf. S.*, 22: 115–142.

Madhupratap, M., 1981. Thermocline and zooplankton distribution. *Ind. J. Mar. Sci.*, 10: 262–265.

Magnesen, T., 1989. Vertical distribution of size fraction in the zooplankton community in Lindarpollene. Western Norway I: Seasonal variation. *Sarsia*, 74: 59–68.

Margalef, R., 1968. *Perspective in Ecological Theory*. Chicago Univ. Press, Chicago, pp. 111.

McAllister, D.E., Schuler, F.W., Roberts, C.M. and Hawkins, J.P., 1994. Mapping and GIS analysis of the global distribution of coral reef fishes on an equal area grid. In: *Mapping the Diversity of Nature*, (Ed.) R.I. Miller. Chapman and Hall Publ., London, p. 155–175.

Mon, T., 1964. *The Pelagic Copepods from Neighbouring Waters of Japan*. The Soyo Co., Tokyo.

Neuman, D., 1989. Circadian components of semilunar and lunar timing mechanisms. *J. Biol. Rhythms*, 4(2): 285–294.

Ohman, M.D., 1990. The demographic benefit of diel vertical migration by zooplankton. *Ecol. Monogr.*, 60: 251–281.

Okemwa. E., 1989. Analysis of six 24 hr series of zooplankton sampling across a tropical creek, the Port Reitz, Mombosa, Kenya. *Trop. Zool.*, 2: 123–128.

Omori, M. and Hamner, W.M., 1982. Patchy distribution of zooplankton: Behaviour, population assessment and sampling problems. *Mar. Biol.*, 72: 193–200.

Omori, M. and Ikeda, T., 1984. *Methods of Marine Zooplankton Ecology*. John Wiley and Sons, New York, p. 86–87, 256.

Paffenhoper, G.A., 1983. Vertical zooplankton distribution on the North-eastern Florida shelf and its relation to temperature and food abundance. *J. Plankton. Res.*, 5(1): 15–33.

Parsons, T.R. and Takahashi, M., 1979. *Biological Oceanographic Processess*, II. Ed. Pergamon Press, Paris.

Peel, R.K., 1974. The measurement of species diversity. *Annual Rev. Ecol. Syst.*, 5: 285–307.

Piatskowski, U., 1985. Distribution, abundance and diurnal migration of acrozooplankton in Antarctic surface waters. *Souderdruckous Bd.*, 4: 64–279.

Pielou, E.C., 1966. Species diversity and pattern of diversity in distribution of ecological succession. *J. Theorit. Biol.* 10: 370–383.

Raymont, J.E.G., 1983. *Plankton and Productivity in the Oceans.* Pergamon Press, Oxford, p. 496–524.

Rothaupt, K.O., 1990. Resource competition of herbivorus zooplankton: A review of approaches and perspective. *Arch. Hydrobiol.*, 118(1): 1–29.

Ryan, T.H., Rodhouse, P.G., Roden, C.M. and Hensey, M.P., 1986. Zooplankton fauna of Killary harbour: The seasonal cycle of abundance. *J. Mar. Biol. Assoc., UK*, 66: 731–749.

Santhanam, R. and Srinivasan, A., 1994. *A Manual of Marine Zooplankton*. Oxford and IBH Publ., Mumbai.

Stich, H.B. and Lampret, W., 1981. Predation evasion as an explanation of diurnal vertical migration by zooplankton. *Nature*, 293(1): 396–398.

Turner, J.T., 1982. The annual cycle of zooplankton in long island estuary. *Estuaries*, 6(4): 261–274.

Valdes, J.L. Roman, M.R., Alvarez Ossorio, M.T., Gauzens, A.L. and Miranda, A., 1990. Zooplankton composition and distribution of the coast of Gallicia, Spain. *J. Plankton. Res.*, 12(3): 629–643.

Vijayalaxmi, 1981. Chaetognatha of Andaman Sea. *Ind. J. Mar. Sci.*, 10(3): 270–213.

Villatte, F., 1991. Annual cycle of zooplankton community in the Abra harbour (Bay of Biscoy): Abundance, composition and size spectra. *J. Plankton Res.*, 13(4): 691–706.

Vinogradov, M.E., 1970. Vertical distribution of the oceanographic zooplankton. Israel Programme for Scientific Translations, Jerusalem, pp. 339.

Washington, H.G., 1989. Diversity, biotic and similarity indices: A review with special relevance to aquatic ecosystem. *Water Res.*, 18(6): 653–694.

Zaret, T.M. and Suffern, J.J., 1976. Vertical migration in zooplankton as a predator avoidance mechanism. *Limnol. Oceanogr.*, 21(6): 804–813.

Zheng, Zhong, 1989. *Marine Planktology*. China Ocean Press, Beijing.

Chapter 4

Biological Control of Locusts and Grasshoppers Using Red-Mite, *Eutrombidium trigonum* Herm.

S.K.A. Rizvi[1], N.P. Singh[2] and Nazima Maqbool[2]
[1]Acridology Laboratory, Al-Noor Plaza, Aligarh, U.P.
[2]Department of Zoology, University Rajasthan, Jaipur, Rajasthan

Introduction

There is no work on this mite which is an important enemy of grasshoppers and locusts (Ref.: COPR, London, 27th April, 1982) and may be a biological control agent for acridid population checks. In India there is no work on this mite except Rizvi, 1974–82, who has reported the occurrence of this mite on *Hieroglyphus nigrorepletus* Bol., on *Patanga succincta* Linn., on *Choreodocus robustus* Ser. etc. He has also some observations on the biological and ecological activities of this parasitic mite with some deleterious effect of this mite on the host. Considering the importance of this mite as an enemy of locusts and grasshoppers it was proposed to investigate the possible biological control of locusts and grasshoppers by using red-mite *Eutrombidium* sp. Locusts and grasshoppers are accepted pests of crops throughout the world. Moreover, keeping in view the pesticide hazards to man and environmental pollution and ecological imbalances it was proposed to intiate work on biological control agents for locusts and grasshoppers with special reference to only *Eutrombidium* sp. And since a comprehensive study on the biology, ecology, population dynamics, host parasite relationship, possibilities of using it as biological agent on commercial basis was already being carried out at A.M.U. Zoological laboratories (Acridology laboratory).

Materials and Methods

Maintenance of stock culture of *Eutrombidium trigonum* Herm. and the Host (*Hieroglyphus, Locusta, Choreodocus, Schistocerca, Acrida, Phlaeoba, Oedaleus, Gastrimargus*).

Above mentioned acridid pests were already in stock in our acridology research laboratory which were reared in wooden cages designed by Hunter Jones (1964) for breeding locusts. The bottom of each cage were provided with metallic oviposition tubes filled with soil. The temperature and relative humidity were maintained at 30±2°C and 70–80 per cent R.H. respectively. Fresh leaves of maize were provided every day as food.

The red-mite, having univoltine life-cycle could only be collected in July and August, 1984 from nearby crop's infested with *Hieroglyphus nigrorepletus* Bol. (a natural host for *Eutrombidium* sp.). These mites were provided, in the laboratory eggs and young nymphs of other acridoids as food.

Observations

These parasitic mites (*Eutrombidium* sp.) appear in field just after the first monsoon shower and then disappears from the host as the winter approaches. These are found not only on the *thoracic* region, as recorded by Peswani (1960), but are also found on the *coxal bases, abdominal tergites* and *sternites, labial bases,* below the *metathoracic extentions* and more so near and around the *external genitalia.*

It was also observed that except first instar hopper, all the nymphal stages and adults of *Hieroglyphus nigrorepletus* Bol. (host) were harbouring mites in the aforesaid body regions. In many samples, the percentage of infestation on the early nymphs, older nymphs and the adults was found to be 19, 27 and 37 per cent respectively. Large number of egg pods of *Hieroglyphus nigrorepletus* Bol. were collected from the fields and examined and thus found to contain two or more mites in an individual egg pod. In twenty cages the egg pods of the grasshoppers were containing more than hundred

(100) mites. The eggs in these pods were found completely damaged while the egg pods of this grasshopper were more harder than those of other acridids. It was also observed that the female grasshoppers were harbouring more mites than male ones.

Peswani (1960) has not accounted for any deleterious effect of these parasitic mite upon the host but we have observed a marked effect of infestation upon the host, adults as well as its eggs. Owing to the presence of this mite, the host *gets highly deleterious* as the mites feed upon the body fluid of the host, eventually *causing death of the host.* It has also been observed that the infested adult host *take more time to mature sexually* with *lesser reproductive potential* than the non-infested one as in case of *Choreodocus illustris* Thunb.

Distribution of Mites

The *Eutrombium trigonum* Herm. (grasshopper mite) a generally distributed in all parts of Aligarh (Lat. 27°34′30″N, 78°4′26″E)

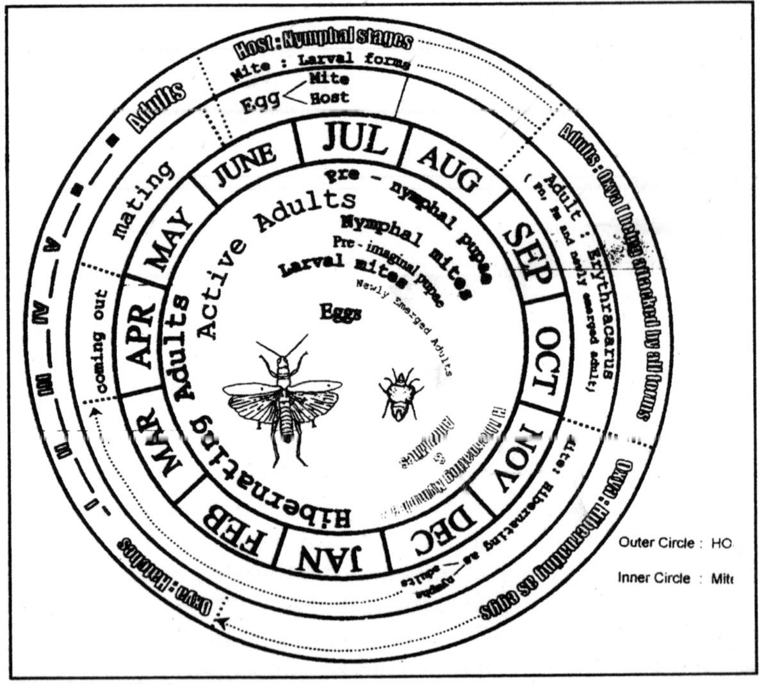

Figure 4.1: Life Stages of Predator and Prey and Synchronization of Biological Profile in an Ecological Docking

particularly in its northern region having good crops and number of wild grasses.

Apparently the mite is not greatly influenced in its distribution by even large variations in rainfall or soil moisture, for it may occur in wet bottom lands as well as on dry patches. Flooding of land for several weeks, however, is damaging to mites and reduce their number materially. The prolonged drought, during which the soil becomes extremely dry, also destroys many of the mites in the late larval, pupal, nymphal and adult stages.

Deleterious Effects

The larval which hatch from the eggs in late June–middle August (a period of acridid outbreaks), are active 6-legged mites. These mites run over the surface and finds grasshoppers to which they attach themselves and obtain their nourishment.

The larvae remain attached to the body of the grasshopper for 10–15 days and engorges itself with the blood of the host.

The following resume on the *deleterious affects* caused by the mites to the grasshoppers based on visual observations and on *certain experiments.*

It has been observed at several places with varieties of grasshoppers and locusts that the larval mite have definite effect upon the health of the grasshoppers by weaken them significantly. The magnitude of effects depends on the severity of infestation of mites and the age of the host. Early instar hoppers are found fallen victim to the larval mite to such an extent that the nymphs of the grasshoppers take longer time in successive moultings, become sluggish and inactive.

Nymphal stages of grasshoppers heavily infested with larval mites have shown *immense deleterious affects* such as *loss in weight, loss in body fluid, crippled developing wing pads, bacterial growth on the punctures made by the mites* and *sometimes death* of the host.

The adult grasshoppers heavily infested with larval mites, are *prevented* from *folding the wings* properly and eventually grasshopper remain unable to fly even to short distances.

The larval mite play a great role in checking the nymphal activities of the young stages of grasshoppers such as retardation in walking activities.

Serious effects on the *hopping behaviour, social aggregation* and *accidental hopper groupings'*, were recorded during several infestation of mites on grasshoppers nymphs and adults. These effects are of immense importance to the acridologists.

Some experiments were designed to find out the effect of mite infestation on the *reproductive potential* of the grasshoppers, data thus obtained are summarised in Table 4.1.

Table 4.1

Sl.No.	*Acridid Pest*	*Normal Number of Egg Laid by a Female*	*Number of Mites per Female*	*Number of Eggs Laid After Parasitization*
1.	*Hieroglyphus nigrorepletus*	60–70	20–30	30–35
2.	*Acrida exallata*	50–60	15–20	25–30
3.	*Phlaeoba infumata*	15–30	10–20	10–15
4.	*Gastrimargus africanus*	20–40	20–30	10–20
5.	*Oedaleus abruptu*	15–20	10–18	10–12
6.	*Choreodocus illustris*	40–50	20–30	25–35
7	*Locusta migratoria*	60–80	30–35	30–40
8.	*Schistocerca gregaria*	65–85	30–35	35–40

Early instar hoppers of grasshoppers (females) with heavy infestation by the mites have shown a rare phenomenon of reproductive arrest to a considerable extent when they moult into an adult female. Parasitic effects by the mite on the nymphal as well as adult female grasshoppers was also considerably significance by preventing them from copulating with their females or making them partially sterile.

Due to the heavy parasitization with the mites, the nymphal stages of the grasshoppers *reduce* their *size* and *retard* their feeding activities and *release watery droplets* from anal opening and seem to be *sick*, mostly try to *aggregate* in *sunny areas* of their places.

The adult and nymphal mites are *predators* of the eggs and body parts of their host (Acridids). It has been recorded that mites prefer to feed upon the liquid contents of grasshopper eggs. *Chorion* is severely collapsed usually die. A nymphal mite may collapse or damage 1–20 grasshopper eggs. An average adult female mite in its pre-oviposition period consume 3–10 grasshopper eggs and may damage 3–25 additional eggs during the remainder of its life.

The consolidated accounts on the predatory potential of the adult and the nymphal mite on the eggs of the grasshoppers is given in Table 4.2.

Table 4.2

Sl.No.	Name of the Pest	Number of Egg Pods in a Glass Jar	Number of Nymphal Mites	Perçentage of Damage to Eggs
1.	*Hieroglyphus nigrorepletus*	10	50	60 per cent
2.	*Acrida exallata*	10	50	60 per cent
3.	*Phlaeobal infumata*	10	50	60 per cent
4.	*Gastrimargus africanus*	10	50	60 per cent
5.	*Oedaleus abruptus*	10	50	60 per cent
6.	*Choreodocus illustris*	10	50	60 per cent
7.	*Locusta migratoria*	10	50	60 per cent
8.	*Scistocerca gregaria*	10	50	60 per cent

Looking at the deleterious effects by the larval mite to the grasshopper nymphs and *predatory potential* of adult and nymphal mites on the eggs of grasshoppers, it can be inferred that the effects are significant and are of practical importance while these mites can be exploited as bio-control agents.

Efficacy and Economic Feasibility as Bio-control Agent

Prior to implementation of acridid control programme a critical appraisal of the practicality of using the agent is essential which requires evaluation of efficacy of the parasite and its economic feasibility as bio-control agent.

After establishing that the *Eutrombidium* sp. is an effective parasite of both, hoppers and adults, stages of locusts and grasshoppers, and predators of their eggs, a series of experiments were conducted to find out the efficacy of its magnitude of the parasitic mite, *Eutrombidium trigonum* Herm. in the integrated control programme against acridid pests. Results thus obtained have shown quite significant efficacy of the parasite on its hosts. This efficacy is less in evaluating the mortality rate but is highly significant in containing the reproductive potential to the extent of sexual sterility. The parasites efficacy was found significant at 5 per cent level.

As predators of the Acridid eggs, the mite's efficacy has been very satisfactory as it produces the population explosion by killing the developing eggs to such extent that the trial areas have shown complete control over next generation population.

The efficacy of the parasite was more pronounced (Laboratory studies) in the retardation of the development of ovarioles in females and in return producing feable number of eggs leading to retardation in the population growth.

Efficacy of the parasite against *Hieroglyphus, Acrida* and *Choreodocus* has been tested while in case of *Locusta, Phlaeoba, Gastrimargus* is yet to be completed when monsoon season will come again.

The efficacy of the parasite against rice grasshoppers, cotton grasshopper and sugarcane grasshopper when tested in the laboratory during July to September 1983 was found considerably significant and productive when the next generation of the same batch of pests were examined during 1984. It was found the efficacy will be more pronounced when this bio-control agent will be used in the integrated control such pests. The significance was at 5 per cent level statistically.

Discussion

Preliminary observations and analysis has revealed that the economic feasibility of this mite (bio-control agent) will be more condusive keeping in view our systems of control operations. Initial investment in the mass production may be visibly more but in due course of time it will be more economic than chemical pesticide. The only difference which has been found out was that of time taken in suppression of grasshopper menace. While bio-control will take more time which will be economically (financially, environmentally, socially) more feasible than chemical pesticides which would be environmentally, financially and socially be more costly than it.

The present investigations on the basis of:

1. Series of experimentation
2. Laboratory trials

3. Mass production of bio-control agents technique
4. Financial implications.

Suggests that the efficacy and the economic feasibility in using this parasitic mite as bio-control agents against locus and grasshoppers would be of economic use to the people engaged in pest management.

Summary

The red-mite, *Eutrombidium trigonum* is an extremely dangerous ectoparasite for live stages and as predators for eggs of grasshoppers and locusts.

Observations have revealed that the larval mites do inflict deleterious effects on the health of the host through loss of body fluid, loss in weight, general weakness and retards all biological activities.

Larval mites, depending on the magnitude of parasitization, have shown direct bearing on the reproductive potentials of the host and retard reproductive gonads to a greater extent.

Adult and nymphal mites are predators of early nymphal instars of the host and produce 50–85 per cent total damage to the eggs of the grasshoppers, thus becoming natural check on grasshoppers/ locusts population and therefore can very well be exploited in the biological control of acridids through such mites.

An effective efficacy of parasitic *Eutrombidium* sp. has been established and it was found that the parasitization leading to deleterious effects caused by this bio-control agent remain significant at 5 per cent level statistically.

Economically the bio-control agent may be more condusive gadget in the integrated control of locusts and grasshoppers.

New Records

1. Deleterious effects of the parasitic mite on the grasshopper stage have been worked out first time in the field of biological control.
2. Eight important grasshoppers and locusts have been recorded susceptible to parasitization of red-mite which was hitherto unknown.

3. Nymphal mites as well as adults have been observed as predators of eggs of the grasshoppers to an extent that a single mite can consume 1–5 or can destroy 1–20 eggs in the egg-pods of the pests.
4. Heavy infestations of larval mites can kill the early instar hoppers of the acridids.
5. Efficacy of the mite as parasitic of acridid pest has been most significant in the suppression of gonad development of the pests.
6. The mite stocking, mass production and its application in the integrated control of acridid pests will be more economic as compared to other means of control methods.

References

Peswani, K.M., 1960. *Eutrombidium trigonum* Hermann (Trombidiidae : Acarina), a predatory red-mite of *Hieroglyphus nigrorepletus* Boliva ("Dhaka"). *Indian J. Ent.*, 22: 236.

Rizvi, S.K.A., 1974. Occurrence of *Eutrombidium trigonum* Herman (Acarina : Trombidiidae) upon *Hieroglyphus nigrorepletus* (Orthoptera Acrididae). *Mushi*, 48: 85–86.

Rizvi, S.K.A., 1975. On the heavy damage caused by earwigs to the egg pods of the Bombay Locust, *Patanga succincta* Linn. in Aligarh breeding grounds. In: *Proc. Nat Acad. Sci.*, held at Bhagalpur. pp. 54 (Abs.) 114.

Rizvi, S.K.A., 1976. Occurrence of *Eutrombidium trigonum* Herm. upon the *Hieroglyphus nigrorepletus* Bol. *Acarologia*, France, 18(1).

Rizvi, S.K.A., 1976. Deleterious effect of the *Eutrombidium trigonum* Herm. upon *Hieroglyphus nigrorepletus* Bol. In: *All India Sympo. Modern Concept on Plant Protection*, held at Udaipur, (Abs.) pp. 39.

Rizvi, S.K.A., 1976. A new record of occurrence of the red-mite, *Eutrombidium trigonum* Herm. on *Patanga succinda* Linn. *Sci. Cult.*, 48: 54–55.

Rizvi, S.K.A., 1978. Life cycle of the parasitic mite, *Eutrombidium trigonum* Hermann (Trombidiidae : Acarina). *Asian Cong. Parasitol.*, Mumbai, pp. 203 .

Rizvi, S.K.A., 1982. A new record of occurrence *Eutrombidium trigonum* Herman, a parasitic red-mite on *Choreodocus robustus* Serville (Orthoptera : Acrididae). In: *Proceeding of the Fourth National Congress of Parasitology,* A.M.U., Aligarh, March 28–30, (Abst.) 32: 61.

Rizvi, S.K.A., 1982. On the field behaviour of the parasitic mite *Eutrombidium trigonum* Herm. (Trombidiidae : Acarina). In: *Eleventh Annual Conference of the Ethological Society of India,* Calicut University, May 3–5, (Abs.) 35: 29.

Rizvi, S.K.A., 1982. On the partial resistance in *Hieroglyphus nigrorepletus* Bol. (Acrididae : Orthoptera) against parasitic mite, *Eutrombidium trigonum* Herm. (Trombidiidae : Acarina). In: *Third Symposium on "Pest Resistance to Pesticide" and 'Synthetic Pyrethroids'* Udaipur.

Rizvi, S.K.A., 1992. On the seasonal abundance of nymphal and adult mites, *Eutrombidium trigonum* Herm. (Eutrombidiinae : Acarina). In: *Fourth National Convention on Indian Society of Life Sciences,* Meerut University, June 12–14, (Abs.) 98: 55.

Rizvi, S.K.A., 1982. On the feeding behaviour of nymphal and adult mite, *Eutrombidium trigonum* Herm. (Eutrombidiinae : Acarina). In: *Fourth National Convention on Indian Society of Life Sciences,* Meerut University. June 12–14, (Abs.) 99: 56.

Rizvi, S.K.A., 1982. Occurrence of the fungus, *Entomophthora grylli* Fresenius on Desert Locust, *Schistocerca gregaria* Forsk. In: *70th Science Congress,* Tirupati.

Rizvi, S.K.A., 1982. Studies on the possible biological control of Desert Locust, *Schistocerca gregaria* Forsk. by the red-mite, *Eutrombidium trigonum* Herm. (Eutrombidiinae : Acarina). In: *70th Science Congress,* Tirupati.

Rizvi, S.K.A., 1985. *An Introduction to Locust.* A.M.U., Aligarh.

Chapter 5

Effect of Different Food Conditions on Survival, Longevity and Life History of *Oithona brevicornis* and *Calanus finmarchicus*

R. Ramanibai and G. Priya*

Biomonitoring and Management Laboratory,
Department of Zoology, University of Madras, Guindy Campus,
Chennai – 600 025, Tamil Nadu, India

ABSTRACT

The cyclopoid and calanoid copepods, *Oithona brevicornis* and *Calanus finmarchicus* when fed with various feeds were monitored in the laboratory. The results of the present study, *O. brevicornis* and *C. finmarchicus* showed highest survival and life span when fed with *Chlorella vulgaris* compared with other feeds. In both species, six naupliar stages (N_1-N_6) and six copepodite stages (C_1-C_6) were observed before attaining the adult stage. The availability of food limitations had severe impact on naupliar and early copepodite stages. The reproduction and development stages showed slight difference between *O. brevicornis* and *C. finmarchicus*.

Keywords: *Copepods, Chlorella vulgaris, developmental stages, life span, survival and rice-bran.*

* *Corresponding Author:* E-mail: rramani8@hotmail.com.

Introduction

Copepods are the major secondary producers in the ocean and play a significant role in the transfer of energy and organic matter from primary producers to higher trophic levels such as planktivorous fishes and carnivorous zooplankton (Parsons *et al.*, 1984). Seasonal variations in terms of body size of copepods have usually been attributed to the effects of temperature and supply of food during growth. Gill and Poulet (1988) investigated the impedance traces of copepod appendage movements illustrating sensory feeding behavior. In both physical and chemical signals to locate and recognize food items, copepods are optimizing their feeding responses and enhancing feeding efficiency. *Oithona* sp. egg production is strongly limited by food during summer and controlled by temperature during winter studied by Sabatini and Kiorboe (1994).

Uye *et al.* (1998) found that the abundance, biomass and production rate of micro and net zooplankton in Dokai inlet, a heavily eutrophic and polluted embayment in northern Kyushu, in August 1996. The amount of food required to support zooplankton secondary production was equivalent to 87 per cent of the phytoplankton primary production, indicating that zooplankton, particularly net zooplankton, are the major phytoplankton grazers in this extremely eutrophic inlet. Laabir *et al.* (2001) determined the ultra-structure of *Temora stylifera* male and female gonads were examined the individuals fed three dinoflagellate and two diatom species which affect copepod hatching success. Zooplankton feeding ecology is a diet of phaeocystis support good copepod grazing survival, egg production and egg hatching success carried out by Turner *et al.* (2002). Jonasdottir *et al.* (2002) studied on the diet composition and quality for *Calanus finmarchicus* egg production and hatching success.

The objective of this study was to pinpoint the survival, longevity and life history of *Oithona brevicornis* and *Calanus finmarchicus*.

Materials and Methods

Study Site

Muttukadu backwater is located 36 km away from Chennai on East Coast Road. Muttukadu backwaters are used for fishing and boating activities. One end of the backwater receives fresh water inflow where as in the other end it joins the sea.

Sampling Method

Zooplankton samples were collected in the month of April 2003, through quantitative oblique hauls were at a depth (1 m) using a 250 µm plankton net. 20 L of surface water was filtered through the net and the planktons were collected in a 100 ml polypropylene containers. Plankton samples were transferred to the laboratory conditions within an hour. *Oithona brevicornis* and *Calanus finmarchicus* were identified and isolated with the help of compound microscope under 100X magnification (Kasturirangan, 1963).

Preparation of Experimental Culture

The culture tanks used for the study were 5-7 L capacity each and washed thoroughly with acid solution (potassium dichromate and conc. H_2SO_4), followed by freshwater and was air-dried. Then the planktons were introduced and allowed to acclimatize to the laboratory conditions for 3-4 days before the commencement of experiments. The culture tank was filled with 5 L of filtered Muttukadu backwater and was well aerated. Cohorts of 20 individuals of *Oithona brevicornis* and *Calanus finmarchicus* were inoculated in two separate tanks. The animals were fed daily with 50 ml of the algae *Chlorella vulgaris* (110 x 10^4cells/ml). The cell densities were determined using a Neubauer haemocytometer. Before inoculating the animals the pH, temperature, salinity and hardness of the culture medium were estimated. The dead animals were discarded periodically. The temperature was maintained at 27±2°C under a photoperiod 12 L : 12 D light-dark cycle.

Experimental Design

Cohorts of 10 adult animals were introduced in 100 ml beakers containing filtered backwaters. *O.brevicornis* and *C.finmarchicus* were fed daily with Unialga *Chlorella vulgaris,* rice bran, yeast and Cow dung. The experiments were carried out in triplicates, fed with *Chlorella vulgaris* at a concentration 114 x 10 cells/ml and lasted for 7 days. The rate of survival and life span of both the experimental organisms were recorded.

Life Table Experiment

Life history experiments were conducted in 100 ml beakers containing 50 ml medium for *O. brevicornis* and *C. finmarchicus* with 3 replicates kept at room temperature (27±2 °C). In each of the beaker, one nauplii was introduced with the help of dissection microscope. Before introducing them into the beaker, the body length of nauplii and copepodites were measured by using occular micrometer under 100X magnification. The medium in each container was changed every day. Life history parameters such as average life span, clutch size, naupliar stage to copepodite stage, copepodite stages to adult stage were monitored under the laboratory.

Results

Survival Rate

The survival rate of *O. brevicornis* and *C. finmarchicus* were shown in Figure 5.1. In both the test species, the mean survival rate

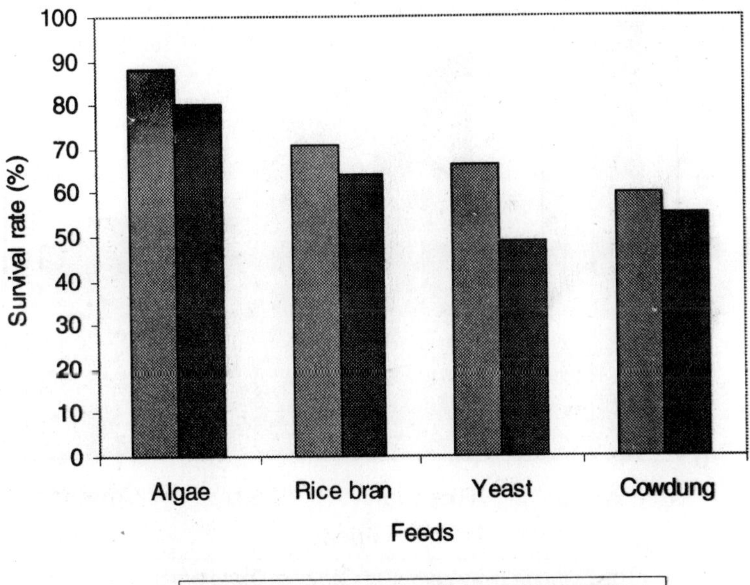

Figure 5.1: Survival Rate of *Oithona brevicornis* and *Calanus finmarchicus* Grown at Different Feeds Under Laboratory Per cent Survival Rate of Individuals Over Four Different Feed

was maximum when fed unialga *Chlorella vulgaris* and recorded as (88.14 per cent, 79.99 per cent), whereas minimum mean survival rate was noted when fed with cowdung (59.85 per cent, 55.14 per cent). When compared with other feeds, ricebran (70.94 per cent, 64.14 per cent) and yeast (66.09 per cent, 48.56 per cent) showed moderate mean survival rate in both the species.

Life Span

Figure 5.2 depicted the life span of *O. brevicornis* and *C. finmarchicus* fed with various food supplements at limited concentration. In *O. brevicornis*, life span was recorded as 31 days when fed *Chlorella vulgaris* whereas *C. finmarchicus* was noted as 26 days. In *O. brevicornis* it was observed to be 29 days whereas *C. finmarchicus* 23 days were noted when fed with rice bran. With yeast as a diet, the life span was recorded as 13 days in *O. brevicornis*.

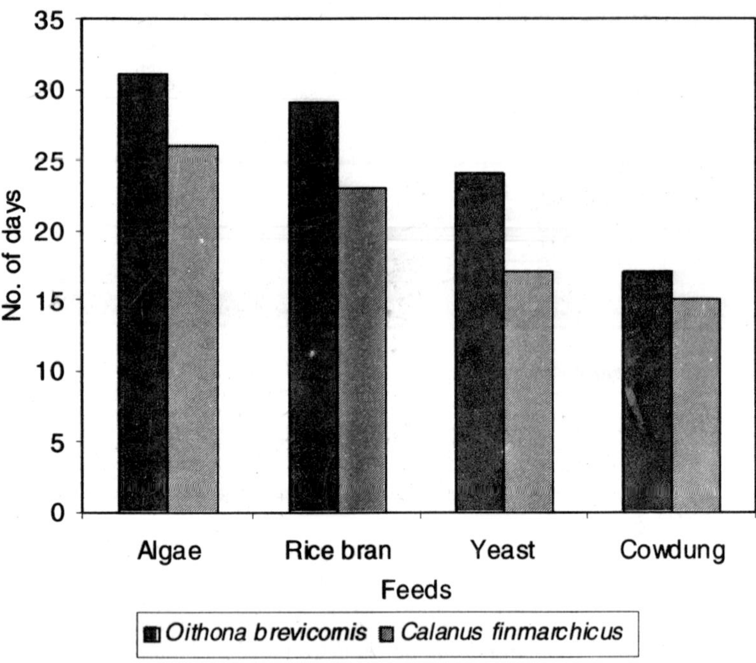

Figure 5.2: Life Span of *Oithona brevicornis* and *Calanus finmarchicus* Exposed to Different Feeds Under Laboratory Conditions

In contrast *C. finmarchicus* was observed 17 days. The life span of *O. brevicornis* was noticed as 17 days whereas *Calanus finmarchicus* it was noted as 15 days when fed with Cow dung. As a concluding remark, both the experimental species survived well when fed with *Chlorella vulgaris*. This indicates that these two species of copepods prefer algal feed for their successful survival.

Life History and Development Stages

The life history of *Oithona brevicornis* was observed from the egg stage through six naupliar stages (N_1–N_6) and 6 copepodite stages (C_1–C_6) before reaching the adult stage. The clutch size of an individual female was 0.18 mm and it released 26 nauplii. The naupliar stages last upto 3 days and copepodite stage for 7 days to become an adult individual. The total life span of *O. brevicornis* was 31 days. Comparatively, life history of *C. finmarchicus* was observed from the egg stage through six naupliar stages (N_1–N_6) and 6 copepodite stages (C_1–C_6) before reaching the adult stage. The egg clutch size of an individual female was 0.1 mm and it released 15 nauplii. The naupliar stages last upto 5 days and copepodite stage for 9 days to become adult individual. The total life span of *Calanus finmarchicus* was 26 days is summarized in Table 5.1.

Table 5.1: Life History Parameters of *Oithona brevicornis* and *Calanus finmarchicus*

Sl.No.	*Life History*	*Oithona brevicornis*	*Calanus finmarchicus*
1.	Size of the animal	0.75 mm	0.98 mm
2.	Size of the clutch	0.18 mm	0.1 mm
3.	Number of nauplii released	26	15
4.	Naupliar stage to copepodite stage (days)	3	5
5.	Copepodite stage to Adult (days)	7	9
6.	Life span (days)	31	26

Discussion

In this study, the abundance of Copepods *Oithona brevicornis* and *Calanus finmarchicus* were assessed in the laboratory. The results showed a slight difference between these two species in terms of life span and survival. The outcome of the present study corresponded with other investigations carried out by Ban *et al.* (1997) showed on reduced hatching rate of *Calanus finmarchicus* eggs in the feeding on the diatom *Thalassiosira nordenskioeldii*, which was one of the dominant diatom species. Symly (1973) found that artifical feeding of *Cyclops abyssorum* on brine shrimp nauplii nearly doubled the mean clutch size and also induced early reproduction. Maier (1990) reported that *Eucylops serrulatus* developed into an adult within 14 days at room temperature. Clutch size of *Oithona brevicornis* generally increases with increasing food level determined by Smyly (1973a, 1973b). Smyly (1980) noted that longevity of adult males and females of *Megacyclops viridis* varied inversely with temperature and that females live longer than males. Postembryonic development level during naupliar and copepodite stages of *Oithona brevicornis* and *Calanus finmarchicus* were increased with increasing salinity ranges. In both species the rate of reproduction and development may be physiologically determined when food is sufficiently abundant. *Oithona brevicornis* and *Calanus finmarchicus*, reproduction and development are related to temperature and are greatly affected by availability of food.

Conclusion

Copepods form the vital link in the food chain that starts from the minute algal cells of the phytoplankton up to large fishes and even mammals. These phytoplankton feeding copepods are the most important primary consumers in marine and freshwater planktonic communities. Copepods form the base of virtually all pelagic food chains. Most commercially exploited fishes in temperate waters feed directly on Copepods during their larval development and they have an essential role to play in the future development of fish farming. Copepods act as pests for many commercially important fishes. Copepods are also important as vectors of some human parasites the most important of these being the guineaworm.

Acknowledgement

One of the authors (R. Ramanibai) thanks the University Grant Commission (UGC), New Delhi, India, for financial support in the form of post doctoral research award.

References

Ban, S., Castel, J., Chandron, Y., Christou, E., Escribano, R., Umani, S.F., Gasparini, S., Uye, S. and Wang, Y., 1997. The paradox of diatom-Copepod interactions. *Mar. Ecol. Prog. Ser.*, 157: 287–293.

Gill, C.W. and Poulet, S.A., 1988. Impedance traces of appendage movements illustrating sensory feeding behaviour. *Hydrobiologia*. 167/168: 303–310.

Jonasdottir, S.H., Gudfinnsson, H.G., Gislason, A. and Astthorsson, O.S., 2002. Diet composition and quality for *Calanus finmarchicus* egg production and hatching success off south-west Iceland. *Marine Biology*, 140: 1195–1206.

Kiorboe, T. and Sabatini, M., 1994. Reproductive and life cycle strategies in egg carrying cyclopoid and free-spowning calanoid copepods. *J. Plankton Res.*, 16: 133–136.

Labbir, M., Buttino, I., Iahora, A., Kattner, G., Poulet, S.A., Romano, G., Carotenuto, Y. and Miralto, A., 2001. Effect of specific dinoflagellate and diatom diets on gamete ultrastructure and fatty acid profiles of the copepod *Temora stylifera*. *Marine Biology*. 138: 124–1250.

Maier, G., 1989. The effect of temperature on the development times of eggs, naupliar and copepodite stages of five species of cyclopoid copepods. *Hydrobiologia*, 184: 79–88.

Parsons, T.R., Takahashi, M. and Hargrave, B., 1984. *Biological Oceanographic Processes*, 3rd edn. Pergamon Press, Oxford, 330 pp.

Smyly, W.J.P., 1973a. Bionamics of *Cyclops strennus, abyssorum sars* (Copepoda, cyclopoida). *Oecologia*, 11: 163–186.

Smyly, W.J.P., 1973b. Decrease in abundance of the freshwater ostracod cypria opthalmica (Jurine) (Crustacea, Ostracoda) in a small lake. *Naturalist, Hull*, 925: 65–69.

Smyly, W.J.P., 1980. Effect of constant and alternating temperatures on adult longevity of the freshwater cyclopoid copepod *Acanthocyclops viridis* (Jurine). *Arch. Hydrobiol.*, 89: 355–362.

Turner, T., Ianora, A., Espostito, F., Carotenuto, Y. and Miralto, A., 2002. Zooplankton feeding ecology does a diet of phaeocystis support good copepod grazing, survival, egg production and egg hatching success. *Journal of Plankton Research*, 24(11): 1185–1195.

Uye, S. and Liang, D., 1998. Copepods attain high abundance, biomass and production in the absence of large predators but suffer connibalistic loss. *Journal of Marine Systems*, 15: 495–501.

Chapter 6

Assessment of Plankton Diversity Using Shannon-Weaver Index (H'): A Case Study in Coleroon and Cauvery Estuaries

P. Elayaraja[1] *and R. Ramanibai*[2]

Biomonitoring and Management Laboratory, Department of Zoology, University of Madras, Guindy Campus, Chennai – 600 025

E-Mail: [1]*elayakal@rediffmail.com;* [2]*rramani8@hotmail.com;*

ABSTRACT

Tamil Nadu coastline extends approximately 1026 km, intersected by several estuarine systems joining the Bay of Bengal, ranging from small water bodies. Coleroon (Tamil = Kollidam) river (11.22°–11.30° N; 79.45°–79.50°E) is one of the major tributaries of river Cauvery and it contributes to the formation of Pichavaram mangroves and wetlands, which are famous for their rich biodiversity and resources. Cauvery on its course to the sea divides into several tributaries and one of which joins the Bay of Bengal at Poompuhar (11.08°N; 79.51°E). A monthly sampling programme was put forth and plankton samples were collected using 55 µm and 20 m mesh nets. The plankton species were identified using standard keys. Enumeration was carried out using the Sedgewick-Rafter counting chamber. Shannon-Weaver diversity index (H') was calculated for phytoplankton and zooplankton separately. The results showed an increase in the species diversity during the post-monsoon season. The index showed low values during the summer season. The phytoplankton index ranged from 0.45 to 2.65 and from 0.23 to 2.45 in Coleroon and Cauvery estuaries respectively. The zooplankton index varied between 0.56 and 2.50 and from 0.56 to 2.56 in Coleroon and Cauvery estuaries respectively. The seasonal variation and abundance of various groups of plankton in these estuaries are discussed in detail in the paper.

***Keywords**: Shannon-Weaver index, Estuary, Plankton, Abundance.*

Introduction

It is not only the high diversity areas that are in need of conservation. It is often in low diversity areas that productivity is highest and human exploit these systems (*e.g.* upwelling areas and estuaries) for food resources and other uses. Estuaries with low species number due to salinity stress are habitats that are under severe threats from urbanization and industrialization. Coastal areas have higher variability of habitats than the open sea and coastal habitats are known to be highly diverse and the greatest threats posed are to these systems. Thus, in the context of biodiversity conservation it is the coastal region that should have the highest priority (GESAMP, 1997).

Biodiversity is a "quality" of biology and ecology rather than a measurable parameter and many indices can be used in an attempt to quantify it. Species are the most practical and widely applicable measure of biodiversity, they are the common currency for biodiversity research and management and the only measure of biodiversity with a well-established standardized code of nomenclature. The relative richness of species in comparable samples can be good indicator of environmental health (Costello, 1998). The most important aspect of biodiversity is the species composition (*i.e.* lists). From checklists of species taken over time (*i.e.*censuses) (Costello, 2000).

Zooplankton is a diverse group of organisms that vary greatly in size and shape, ranging from millimeter sized crustaceans to jellyfish larger then a meter in diameter. Zooplankton are ubiquitous and abundant in most marine habitats and plays an important role in many pelagic ecosystems. They are important grazers of phytoplankton and are a significant portion of the diet of many larval and juvenile fish. The field of zooplankton ecology involves studying the relationship between zooplankton and the physical and biological factors affecting them or being influence by them. Phytoplankton productivity contributes about 95 per cent of total production in the marine environment. Hence the abundance of phytoplankton can be taken as the best means of measuring primary production. Also some phytoplankton species are often used as good indicators of water quality including pollution. During the past, extensive work has been done pertaining to qualitative and quantitative ecology of phytoplankton from both the Indian coasts (Gouda and Panigrahy, 1996).

It has been observed that coastal areas with large estuaries tend to support high fishery yields. By providing abundant food and shelter, estuaries are utilized by many species and their most important role is the provision of nursery grounds for juvenile of certain marine species, many of which are directly or indirectly of commercial importance.

Several studies have shown that when comparing zooplankton communities in a particular geographical zone, some zooplankton assemblages can be found to characterize different types of water bodies (Armengol and Miracle, 1999). Various species data at large scales are increasingly documented in this presence/absence format (Beaugrand et al., 2002). Shannon-Weiner Diversity (H) is a function of both the number of species in a sample and the distribution of individuals among those species. Low H is often a result of poor species richness, dominance by one or more species, or a combination of both.

Materials and Methods

Two estuaries, Coleroon and Cauvery along the Bay of Bengal were selected for the present study. The Coleroon (Tamil = Kollidam) river (11.22 °–11.30 ° N; 79.45°–79.50° E) is one of the major tributaries of river Cauvery. It serves as a major source of irrigation in the area. The study of this estuary assumes importance in the context of pesticide pollution and nutrient loading from the land drainage. The river joins the Bay of Bengal through a narrow mouth at Mahendrapalli, near Kollidam town. The Coleroon river also contributes to the formation of Pichavaram mangroves and wetlands, which are famous for its rich biodiversity and resources.

Cauvery on its course to the sea divides into several tributaries. One of which joins the Bay of Bengal at Poompuhar (11.08°N; 79.51°E). For the past few years the mouth of the estuaries has remained closed during the summer season due to severe drought. The effect of agriculture and aquaculture on the estuarine planktonic diversity is worth examining.

The monthly sampling lasted for 22 months from July 2002 to April 2005. Plankton samples were collected filtering 100 L of surface water through a 200 µm and 50 µm nylon mesh nets for zooplankton and phytoplankton collection respectively. The samples were preserved in 5 per cent sea water buffered formalin and analyzed in the laboratory under a Carl-Zeiss compound microscope. Plankton counting was carried out using a Sedgewick-Rafter counting chamber and identification of species was done using standard identification keys (Desikachary, 1959; Edmondson, 1959; Kasthurirangan, 1963; Subrahmanian, 1964; Reddy and Radhakrishna, 1984; Anand, 1989; Battish, 1992; Anand, 1998).

Shannon-Weaver diversity index (*H'*) was calculated using the following formula (Shannon and Weaver, 1963).

$$H' = \sum_{i=1}^{n} pi \log pi$$

The values were cross-checked using software packages like PRIMER v5 (demo version) obtained from Plymouth Marine Laboratory (PML), UK.

Results

About 65 species of zooplankton and 105 species of phytoplankton were recorded from Coleroon and Cauvery estuaries during the study.

The Shannon-Weaver index for zooplankton diversity varied between 0.56 to 2.5 in Coleroon and 0.56 to 2.6 in Cauvery estuary. The phytoplankton index varied between 0.45 to 2.62 in Coleroon and 0.24 to 2.4 in Cauvery estuary.

Figure 6.1 shows the monthly variation of Shannon index of zooplankton and phytoplankton in Coleroon estuary. There seems

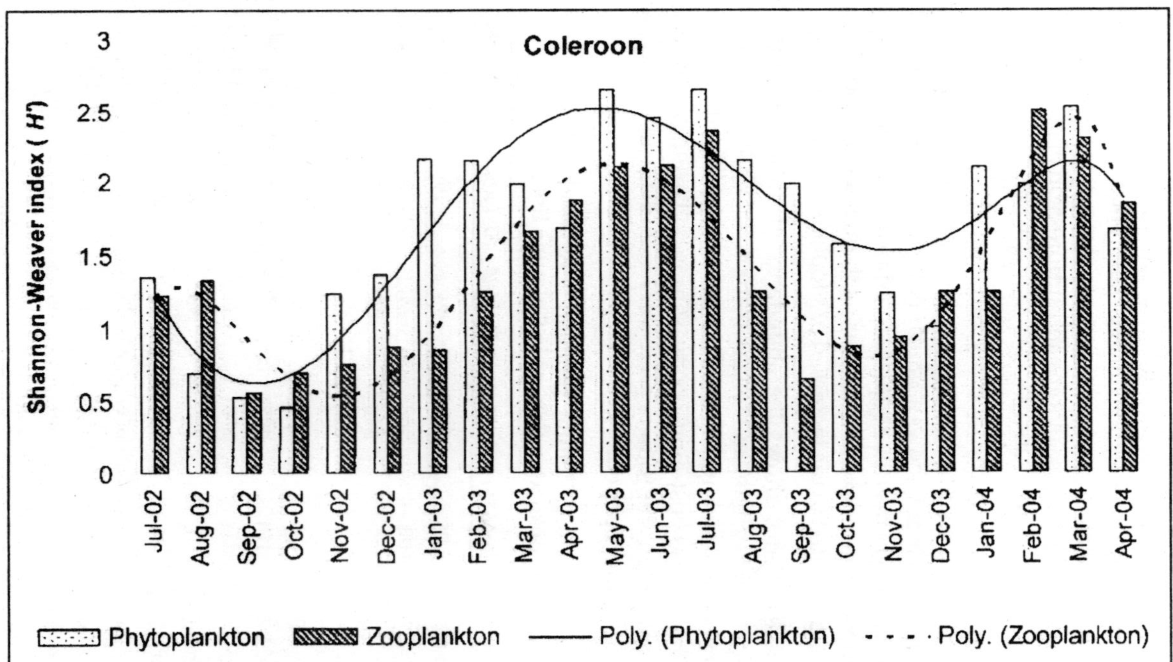

Figure 6.1: Monthly Variation of Zooplankton and Phytoplankton Diversity in Coleroon Estuary

to be a positive relationship between zooplankton and phytoplankton. The peaks are observed during the pre-monsoon months and post monsoon months, whereas the lowest diversity was recorded during the monsoon seasons. This could be because of disproportionate growth of certain algal species (especially the blue-greens) which cause blooms and suppression of other species and thus decrease in diversity.

Figure 6.2 shows the monthly variation of Shannon index of zooplankton and phytoplankton in Cauvery estuary. The phytoplankton index shows a similar pattern that was observed in the Coleroon, whereas the zooplankton showed a different pattern. The zooplankton index increased during the monsoon seasons. This could be attributed to the increased phytoplankton production.

Tables 6.1–6.4 shows the abundance of various groups of zooplankton and phytoplankton in Coleroon and Cauvery estuaries. The increase in the abundance of blue-green algae is evident during the monsoon and post-monsoon seasons in both estuaries. Among the zooplankton the rotifers were very dominant in abundance followed by the copepods.

Table 6.1: Monthly Variation of Abundance of Zooplankton Groups in Coleroon Estuary

Month	*Copepoda*	*Rotifera*	*Tintinnid*	*Ostracoda*	*Cladocera*
Jul-02	125	668	46	16	12
Aug-02	68	562	25	24	5
Sep-02	65	896	65	25	16
Oct-02	98	135	48	23	35
Nov-02	126	689	125	13	26
Dec-02	154	465	133	32	45
Jan-03	213	356	125	16	12
Feb-03	102	246	98	25	11
Mar-03	89	132	75	21	2
Apr-03	78	656	65	25	11
May-03	99	152	46	22	23
Jun-03	65	621	85	11	6
Jul-03	49	326	89	22	23
Aug-03	19	698	35	22	4

Contd...

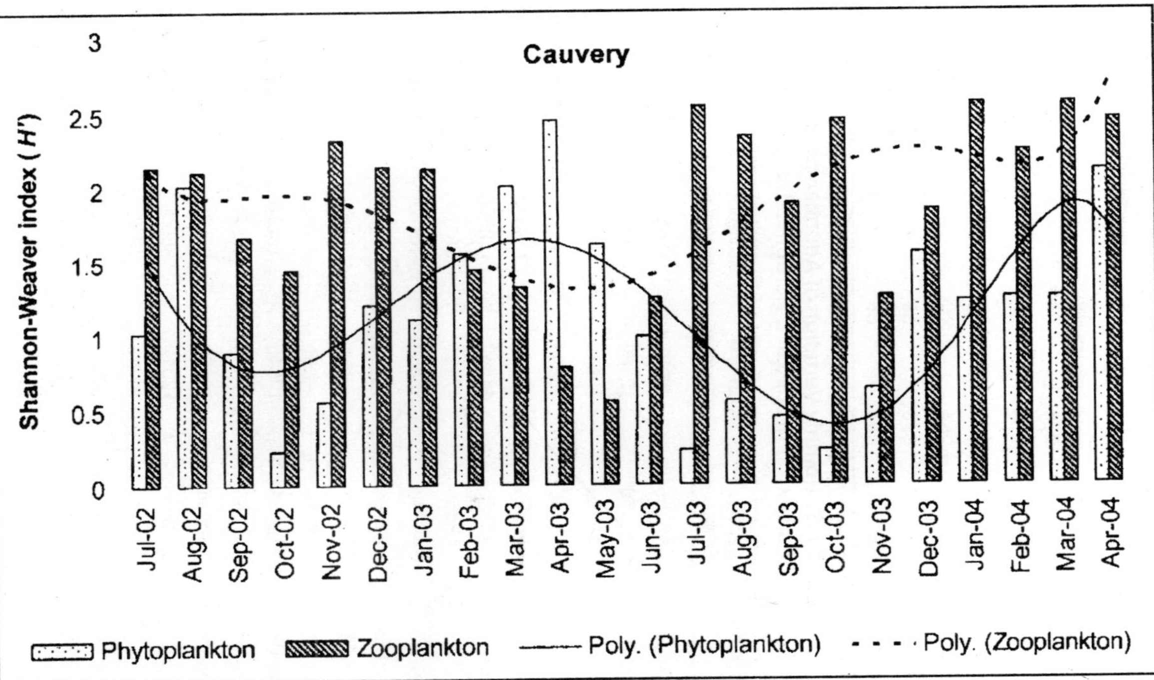

Figure 6.2: Monthly Variation of Zooplankton and Phytoplankton Diversity in Cauvery Estuary

Table 6.1–Contd...

Month	*Copepoda*	*Rotifera*	*Tintinnid*	*Ostracoda*	*Cladocera*
Sep-03	46	595	65	55	22
Oct-03	46	425	13	22	42
Nov-03	99	366	65	22	35
Dec-03	56	156	36	11	12
Jan-04	49	166	66	34	16
Feb-04	68	254	213	42	16
Mar-04	26	198	65	26	3
Apr-04	16	295	65	35	3

Table 6.2: Monthly Variation of Abundance of Zooplankton Groups in Coleroon Estuary

Month	*Bacillario-phyceae*	*Chloro-phyceae*	*Cyano-phyceae*	*Dyno-phyceae*
Jul-02	16356	8569	126563	58
Aug-02	23566	12536	96533	76
Sep-02	45266	12546	36563	95
Oct-02	45245	26544	46532	86
Nov-02	56224	39564	124562	99
Dec-02	42566	48562	152566	65
Jan-03	35666	47525	98564	75
Feb-03	45666	52654	16463	42
Mar-03	24564	24562	36855	66
Apr-03	23556	9564	26566	23
May-03	15244	7256	85656	66
Jun-03	12564	12545	16556	54
Jul-03	11245	12452	46596	14
Aug-03	32656	32562	65665	55
Sep-03	7859	12012	85656	94
Oct-03	14022	9546	46466	69
Nov-03	16524	8256	156324	97
Dec-03	13255	7566	132456	66
Jan-04	65244	9564	112456	26
Feb-04	4655	12564	98333	35
Mar-04	32656	6552	65232	56
Apr-04	3265	4256	46466	95

Table 6.3: Monthly Variation of Abundance of Zooplankton Groups in Cauvery Estuary

Month	*Copepoda*	*Rotifera*	*Tintinnid*	*Ostracoda*	*Cladocera*
Jul-02	56	78	55	24	25
Aug-02	35	659	22	33	33
Sep-02	24	598	43	45	65
Oct-02	52	398	66	21	42
Nov-02	46	456	46	54	89
Dec-02	75	689	24	33	85
Jan-03	52	465	53	26	75
Feb-03	42	235	52	22	62
Mar-03	16	356	15	33	45
Apr-03	36	326	25	24	21
May-03	25	755	46	26	0
Jun-03	44	256	32	35	0
Jul-03	48	135	21	12	15
Aug-03	31	332	55	24	10
Sep-03	36	456	20	44	12
Oct-03	89	166	52	22	15
Nov-03	85	699	40	55	58
Dec-03	69	565	52	11	66
Jan-04	55	326	23	16	46
Feb-04	22	232	26	25	15
Mar-04	35	323	36	66	2
Apr-04	25	466	21	55	0

Table 6.4: Monthly Variation of Abundance of Zooplankton Groups in Cauvery Estuary

Month	*Bacillario-phyceae*	*Chloro-phyceae*	*Cyano-phyceae*	*Dyno-phyceae*
Jul-02	156000	526	196225	465
Aug-02	175800	4585	256544	698
Sep-02	152000	566	89526	1025
Oct-02	123000	522	336666	2452
Nov-02	89551	566	266999	1356

Contd...

Table 6.4–Contd...

Month	*Bacillario-phyceae*	*Chloro-phyceae*	*Cyano-phyceae*	*Dyno-phyceae*
Dec-02	75235	8659	63555	1100
Jan-03	65888	5444	89522	2451
Feb-03	48955	265	66333	568
Mar-03	45688	785	45855	846
Apr-03	45855	4999	15665	899
May-03	45522	1205	24556	2422
Jun-03	23633	222	26632	1025
Jul-03	16996	1155	56332	266
Aug-03	204653	236	89566	245
Sep-03	189255	266	156623	2541
Oct-03	120255	6555	15545	555
Nov-03	165488	2566	2333	1222
Dec-03	80522	2336	63656	1025
Jan-04	36444	552	358996	988
Feb-04	56966	269	42566	662
Mar-04	145555	4522	68566	1466
Apr-04	195455	2666	45599	785

Acknowledgements

One of the Authors (P.E) would like to thank the Council of Scientific and Industrial Research (CSIR) for the financial grant in the form of Junior Research Fellowship (JRF-NET) and Senior Research Fellowship (SRF).

References

Anand, N., 1989. *Handbook of Bluegreen Algae*. Bishen singh Mahendra Pal Singh, Dehradun, India.

Anand, N., 1998. *Indian Freshwater Microalgae*. Bishen singh Mahendra Pal Singh, Dehradun, India.

Armengol, X. and Miracle, M.R., 1999. Zooplankton communities in doline lakes and pools, in relation to some bathymetric parameters and physical and chemical variables. *J. Plankton Res.*, 21: 2245–2261.

Battish, S.K., 1992. *Freshwater Zooplankton of India*. Oxford and IBH Publishing Co. Pvt. Ltd., New Delhi, India.

Beaugrand, G., Ibanez, F., Lindley, J.A. and Reid, P.C., 2002. Diversity of calanoid copepods in the North Atlantic and adjacent seas: species associated and biogeography. *Mar. Ecol. Prog. Ser.*, 232: 179–195.

Costello, M.J., 1998. To know, research, manage and conserve marine biodiversity. *Oceanis*, 24: 25–49.

Costello, M.J., 2000. Developing species information system: The European Register of Marine Species (ERMS). *Oceanography*, 13: 48–55.

Desikachary, T.V., 1959. *Cyanophyta*. ICAR, New Delhi, pp. 686.

Edmondson, W.T., 1959. *Freshwater Biology*, 2nd edn. John Wiley and Sons, NY, USA.

GESAMP (IMO/FAO/UNESCO–IOC/WMO/WHO/IAEA/UN/UNEP. Joint Group of Experts on the Scientific Aspects of Marine Environmental Protection), 1997. Marine Biodiversity: patterns, threats and conservation needs. Rep. Stud. GESAMP, 62: 24p.

Gouda, R. and Panigrahy, R.C., 1996. Ecology of phytoplankton in coastal waters off Gopalpur, east coast of India. *Ind. J. Mar. Sci.*, 25: 81–84.

Kasthurirangan, C.R., 1963. *A Key for the Identification of the more Common Planktonic Copepods of Indian Coastal Waters*. CSIR, New Delhi, India.

Reddy, R.Y. and Radhakrishna, Y., 1984. The calanoid and cyclopoid fauna (crustacean, copepoda) of lake Kollaru, South India. *Hydrobiologia*, 119: 27–48.

Shannon, C.E. and Weaver, W., 1963. *The Mathematical Theory of Communication*. University of Illinois Press, Urbana.

Subrahmanian, K., 1964. Diatoms of Madras coast. In: *The Proceedings of the Indian Academy of Sciences*. 24(4) Sec. B.

Chapter 7

Ecology of Aquatic Insects in Harike Wetland

K.S. Bath

Senior Scientific Officer,

Punjab State Council for Science and Technology, Chandigarh

Aquatic insects are those that spend some part of their life-cycle closely associated with water, either living beneath the surface or skimming along on top of the water. Aquatic insects are among the most prolific animals on earth, but are highly specialized and represent less than 1 per cent of the total animal diversity (Pennak 1978). A lot of work on the taxonomy, habitats, communities, life history, adaptations, trophic relationships, predation, sampling techniques etc. of aquatic insects has been carried out by many workers in India and abroad. Similarly, correlation studies of aquatic insects with physico-chemical parameters and other animal groups have also been studied widely. Insects have also been used as bio-indicators in various aquatic ecosystems because aquatic insects are excellent overall indicators of both recent and long-term environmental conditions (Patrick and Palavage 1994). The immature stages of aquatic insects have short life cycles, often several generations a year, and remain in the general area of propagation. Thus, when environmental changes occur, the species must endure the disturbance, adapt quickly, or die and be replaced by more tolerant species. These changes often result in an overabundance of a few tolerant species, and the communities become destabilized or "unbalanced."

Harike wetland is one of the three Ramsar sites in Punjab which nurtures a wide variety of flora and fauna. This internationally recognized wetland is located around 31° 13′ N Latitude and 75°12′ E Longitude. Harike attracts large populations of avifauna, in particular diving ducks. The scaup duck, falcated teal and the whiteheaded stifftailed duck are species rarely seen elsewhere within India. Harike is a vital source of water and fish for the people of Punjab. It supports rare, vulnerable and endangered faunal species, which include the testudine turtle and the smooth Indian otter, both of which are listed in the IUCN list of threatened animals. Over 20,000 ducks have been recorded here during the peak migratory season. The *wigeon, common teal, pintail, shoveller and brahminy* ducks are commonly seen during the winter. The lake is particularly famous for diving ducks, such as the crested pochard, common pochard and tufted ducks which occur in very large numbers. A total of 167 bird species were recorded during 1980-85 period; 40 species were long distance migrants, which pass through, or winter at Harike lake. Apart from avifauna, some 7 species of turtle and 26 species of fish have been recorded. The mammals found at Harike include the smooth Indian otter, the jungle cat, jackal, Indian wild boar and the common mongoose.

Various sites of Harike wetland were subjected to extensive limnological investigations to include physical, chemical and biological parameters during 1990-1993 under a *Punjab State Council for Science and Technology* sponsored project titled *'Analytical studies of Aquatic Ecosystem of Punjab'* by a team of researchers from *Department of Zoology, Punjabi University, Patiala.*

Aquatic Insects: The aquatic insects encountered at Harike belonged to the orders *Odonata, Coleoptera, Diptera, Ephemeroptera, Hemiptera, Neuroptera, Trichoptera* and *Hymenoptera.* These orders were represented by 32 taxa (Bath, 1997).

The ecological studies conducted at various sites indicated that sites that harboured rich growth of aquatic vegetation had more insect population because vegetation provides shelter, protection and food to the insects. Seasonal variation studies depicted that population of insects was maximum during the summer months of April to October (except during the months of heavy rains) when the temperature ranged between 30°C and 40°C. During this season, the insect population rose twice.

Table 7.1: Identified Aquatic Insects at Harike (1990-1993)

Order and Taxa	*Order and Taxa*
Odonata:	**Diptera:**
Aeshna sp.	*Simulium* sp.
Heganius sp.	*Culex* sp.
Anax sp.	*Anopheles* sp.
Coleoptera:	*Chironomus* sp.
Dytiscus sp.	*Dixa* sp.
Hydrophilus sp.	*Phantom* sp.
Dryops sp.	*Chaoborus* sp.
Hemiptera:	*Forcipoymea* sp.
Deronectus sp.	*Dicranota* sp.
Hydrometra sp.	*Tanypus* sp.
Plea leachi	*Pentaneura* sp.
Notonecta sp.	**Trichoptera:**
Ephemeroptera	*Leptocorus* sp.
Ephemerella sp.	*Phryganea* sp.
Baetis sp.	*Caraclea sp*
Hymenoptera:	
Polynema natans	

Bio-indicator species studies at Harike labeled dipteran taxa of *Simulium* sp., *Culex* sp., *Chironomus* sp. *and Dixa* sp. as tolerant insect taxa, found in a wide range of physico-chemical environments. Some other taxa like *Phantom* sp., *Chaoborus* sp., *Tanypus* sp. *and Pentaneura* sp. were found inhabiting only polluted sites. These taxa were labeled as indicators of industrial and organic pollution.

Correlation studies conducted at Harike indicated that insect populations grew well when the values of temperature, fluorides and total hardness were high. However, rising chloride and phosphate values had an adverse effect on aquatic insects.

As in the case of other wetlands, Harike is also threatened by a number of factors. The major causes of wetland loss and degradation include: Large scale utilization of both surface and ground waters for irrigation; The expansion of intensive agriculture resulting in encroachments on the wetland habitat; Drainage of agricultural chemicals into the waters; Discharge of untreated waste from catchment towns into the rivers which feed the wetland, and consequent weed spread; soil erosion and siltation. Apart from the conservation and management measures being undertaken, there is an urgent need to regularly conduct detailed ecological studies on insects and other groups so as to define their importance and role in the delicate aquatic food web.

References

Bath, K.S., 1997. Limnological studies of the Harike wetland ecosystem. *Ph.D. Thesis,* Department of Zoology, Punjabi University, Patiala.

Pennak, R.W., 1978. *Freshwater Invertebrates of the United States,* 2nd edn. John Wiley and Sons, Inc., New York, 803 pp.

Patrick, R., and Palavage, D.M., 1994. The value of species as indicators of water quality. In: *Proceedings of the Academy of Natural Sciences,* Philadelphia, 145: 55–92.

Chapter 8

Ecology of Kanjli Wetland with Reference to the Invertebrate Fauna

K.S. Bath

Senior Scientific Officer,

Punjab State Council for Science and Technology, Chandigarh

Wetlands are unique ecosystems. These are important on account of their hydro-biological, ecological, socio-cultural, religious and recreational values. They play an important and crucial role in the journey of water from oceans to atmosphere through wetlands (rivers, streams, lakes, ponds), land and back to the oceans in the water cycle. Wetlands also play a vital role in retaining initial enhanced flows in rivers from melting ice caps due to global warming and thus protecting from floods. They support unique biodiversity, deliver goods and services for the mankind and provide spiritual solace to the wavering human minds, help sustenance of associated ecosystems and reminds us about the natures bounties. Their association with the mankind is ancient as civilization thrived along the wetland areas. They provided the man with all basic necessities starting the water for drinking and irrigation of his cropland areas to the supply of food and construction material for the households. Their association with the mankind is ancient as civilization thrived along the wetland areas. Even today wetlands provide innumerable benefits directly or indirectly to the human beings.

Punjab has around 42 wetlands covering of a total area of around 23000 ha. Three wetlands namely Harike, Ropar and Kanjli have been included in the Ramsar list of wetlands of International importance. Kanjli Wetland named after the village Kanjli is at a distance of 4 km from Kapurthala town. Geographically it is situated at 31' 25'–31' 28' North Latitude and 75' 23'–75' 25' East Longitude at an elevation of 210 meters above Mean Sea Level. It falls under the administrative boundary of Kapurthala district of Punjab. Kali Bein rivulet on which Kanjli Wetland is located used to be one of the important tributaries (live branch) of river Beas. It has, of late, become independent of river Beas due to silting up of the Bein and westward shifting of Beas. The Bein travels a long distance after originating from near village Dhanoa a few kilometers upstream of Budho Barkat Regulator in Hoshiarpur District and feeds the Kanjli lake and the wetland areas. It further moves towards Bakarke village, 10 kms. short of Harike Pattan Regulator and joins river Beas. The wetland is located about 75 kms North-East of Harike. It is a permanent fresh water stream converted into a small reservoir at Kanjli for the purpose of irrigation supplies. Depth of water varies from 10 feet to 25 feet depending upon the season and water inflow. Catchment area is mainly under agriculture. Kali Bein is a permanent rivulet. The Bein in general and Kanjli Lake in particular act as important water sponge water discharging and its recharging. Excess water during rainy seasons from the adjoining agricultural crops gets discharged into the Kali Bein. Its importance as a discharging drain becomes more significant when it helps by taking away excessive rainwater from sensitive crops like wheat, potato, etc. It also serves as an important source for agriculture. At the same time it also helps in flood management. The ground water is fast receding in many districts of Punjab. Year after year the dugwells as well as tubewells are deepened to take out water for irrigation. But the observations around Kanjli wetland show that the ground water table is stable as compared to rest of the state. The Bein is thus playing a vital role in enriching the ground water thereby helping the nearby population, which is dependent on ground water for irrigation, industries and drinking supplies, for easy abstraction of water. Pressure on underground water is also somewhat relieved as a number of farmers do direct abstraction of water from the Kali Bein, as it is very economical.

Ecological Values

Ecologically Kanjli Wetland has similar values like Harike Wetland. It supports diversity of life forms. 26 ha forest area created along the Kali Bein provide suitable habitat for various organisms. Various species of mammals, fishes, birds, other vertebrate fauna, microfauna and large species populations of plants are important features of this wetland.

Kanjli Wetland support deep water, shallow water, mesic as well as xeric features. The food chain is basically of grazing type. The abiotic components of this ecosystem are sun light, pH, inorganic salts, nutrients and dissolved gases. The organic matter accumulates at the bottom and mainly comes from the death and decay of plants and animals. Producers are autotrophic plants and some photosynthetic bacteria. These are mainly the rooted submerged, floating and emergent hydrophytes like *Typha* sp., *Eleocharis* sp., *Sagittaria* sp., *Nymphaea* sp., *Potamogeton* sp., *Vallisnaria* sp., *Eichhornia* sp., *Lemna* sp. *etc.* and minute, floating or suspended lower plants like filamentous algae, diatoms, chlorococcales and flagellates. The primary consumers at Kanjli are herbivores. These include mollusks, crustaceans, rotifers, some insects and fish. Detrivores feed upon plant remains and organic matter. Besides, some mammals such as buffaloes, cows, etc. also visit the lake and feed on marginal rooted macrophytes. Some birds also feed on some hydrophytes. The secondary consumers are carnivores, feeding on insects, mollusks, rotifers and crustaceans. Some insects and carnivorous fishes feed on crustaceans, rotifers and mollusks. The tertiary consumers are some snakes and birds which feed on small fishes and insects. The upper most consumer level is occupied by man and fish eating and scavenger birds. The organic matter which accumulates at the bottom of the lake and forested wetland area is decomposed by a variety of heterotrophic mircrobes-bacteria and fungi.

Invertebrate Fauna at Kanjli Wetland

Studies carried out by NEERI, Nagpur and Deptt. of Zoology, Punjabi University shows the availability of diverse kind of zooplankton species at Kanjli Wetland. The dominant zooplankton species include *Diffulgia* sp., *Vorticella* sp., *Brachionus* sp., *Epiphanes* sp., *Monostyla* sp., *Testudinella* sp., *Chaetogaster* sp., *Diaphanosoma* sp., *Ceriodaphnia* sp., *Chydorus* sp., *Macrothrix* sp., *Cyclops* sp.,

Mesocyclops leuckartii and *Chiromomus larvae.* However, their occurrence varies from season to season. The Punjab Pollution Control Board (PPCB) Patiala (1993) has reported the dominant presence of some more species like *Notholca* sp., *Daphnia* sp., *Diaptomas* sp., *Nauplius* sp., *Microstomum* sp. and *Bothromesostomum* sp. Similarly the Benthic Macroinvertebrate species reported in Kanjli Wetland area includes *Branchiura sawerbyii, Baetis* sp., *Chironomus tentans* and *Chironomus tendipediformis*. The PPCB has also reported the dominant presence of *Planorbis* sp. *Heteropterasp. and Chironomids.* The dominant insects encountered at Kanjli include *Hydropsyche* sp., *Limephilus* sp., *Baetis* sp., *Hexagenia limbata, Tanypus* sp., *Palpomya* sp., *Chironomus tentans* and *Chironomus tendipedifomis*.

Table 8.1: Invertebrate Fauna of Kanjli Wetland

Zooplankton Species	
Protozoa:	**Cladocera:**
Amoeba sp.	*Alona* sp.
Cemtrpjxis sp.	*Bosmina* sp.
Coleps hirtus	*Ceriodaphnia* sp.
Diffugia sp.	*Chydorus* sp.
Vorticella sp.	*Pleuroxus* sp.
Rotifera:	*Macrothrix* sp.
Anuroeposis sp.	**Copepoda:**
Brachionus sp.	*Cyclops* sp.
Epiphanes sp.	*Mesocyclops leuckarti*
Filinia longiseta	**Benthic macroinvertebrate species**
Keratella valga	**Insecta:**
Lecane sp.	*Hydropsyche* sp.
Lepadella sp.	*Limnephilus* sp.
Monostyla s.	*Baetis* sp.
Platyias sp.	*Hexagenia limbata*
Ploesoma sp.	*Tanypus* sp.
Testudinella sp.	*Palpomya* sp.
Trichotria sp.	*Chironomus tentans*
	Chironomus tendipediforms

Contd...

Table 8.1–Contd...

Zooplankton Species	
Namatoda:	**Oligochaeta:**
Monochulus sp.	*Aelosoma bengalensis*
Monochus sp.	*Chaetogaster* sp.
Rhabdolaimus sp.	*Branchiura sawerbyii*
Oligochaeta:	**Mollusca:**
Chaetogaster sp.	*Lamellidan marginalis*
Diaphanosoma sp.	*Indoplanorbis exustus*
Diplogaster sp.	*Lymnaea lutiola*
	Melanoides sp.

Due to heavy pressure on the wetland areas on account of various factors like encroachment for agriculture, the impact on faunal populations is catastrophic. It is clear that unless solid argument based on hard scientific data is presented for maintenance of these sites, this effort is likely to continue. Hence the need of investigation in this area. Detailed taxonomic studies of plant and animal species of this wetland need to be carried out. This will also help to identify the endemic species, if any, of this region that will invite particular attention for conservation. Studies of fish biology and biodiversity conservation, aquaculture, socio-economics, database establishment and fish stock assessment will be useful.

References

Handa, B.K., 1993. *Management and Control of Aquatic Weeds in Kanjli Lake,* District Kapurthala (Punjab). NEERI, Nagpur.

Punjab Pollution Control Board, 1999. *Kanjli Wetland: A Report on Status of Water Quality,* Patiala.

Punjab State Council for Science and Technology, 1998. *Comprehensive Conservation and Management Plan (for Ninth Five Year Plan) for Kanjli Wetland.*

Punjab State Council for Science and Technology, 2000. *Integrated Plan for Conservation and Management of Kanjli Wetland.*

Punjab State Council for Science and Technology, 2003. *Kanjli Wetland.*

Chapter 9

Distribution of Insects in Relation to Pollution in Satluj River, Punjab

H. Kaur, J. Syal, S. Jasuja and S.S. Dhillon*
Department of Zoology, Punjabi University, Patiala – 147 002

ABSTRACT

River Satluj, the main life line of Punjab state, originates from Mansarovar lake in Tibet, enters Punjab at Nangal, flows through districts of Ropar, Ludhiana, Jalandhar and finally joins the river Beas at Harike. During its course, it gets heavily polluted by domestic sewage, agricultural run-off and untreated or inadequately treated effluents discharged from various types of industrial units which affect the physico -chemical as well as biological status of the river as is revealed by a detailed analytical study. The present paper includes studies on species composition of insect fauna in various sections of the river. In all, twenty species of insects have been recorded with 6 at site-I, 13 at site-II, 18 at site-III, 7 at site-IV, 5 at site-V, 5 at site-VI, 4 at site-VII, 2 at site-VIII, 7 at site-IX and 8 at site-X. The study reveals maximum species diversity at the site-III where water is fairly slow and is not much polluted whereas at the polluted sites of the river the number of insect species is fairly low.

***Keywords**: Insects, Satluj river, Species diversity.*

---

* *Corresponding Author:* E-mail: harbhajankaur@hotmail.com.

Introduction

River Satluj originates from Mansarovar lake in Tibet. It enters Punjab at Nangal, flows through Ropar, Ludhiana, Jalandhar districts and finally joins river Beas at Harike after which it leaves Indian territory next to Ferozpur to enter Pakistan. The river Satluj, one of the main rivers of Punjab state, provides water for irrigation and fulfills drinking needs. Due to urbanization and industrialization, it is being polluted by domestic and industrial effluents coming from human settlements and various types of industries thereby changing its ecological conditions. The present study includes distribution of insect fauna in relation to pollution in various regions of the Satluj river as it courses through the Punjab state.

Material and Methods

Selection of Sites

Ten sites which were selected along the course of river Satluj keeping in mind topographical features are shown in the map and the pollution sources at each site are as follows:

Table 9.1: Showing Pollution Sources of Ten Selected Sites Along the Course of River Satluj

Sl.No.	*Sites*	*Pollution Source*
1.	Site-I (Upstream of Nangal reservoir)	No point-source pollution
2.	Site-II (Downstream of Nangal reservoir)	Wastes from PNFC, NFL, PACL
3.	Site-III (Anandpur Sahib)	No fresh point-source pollution
4.	Site-IV (Kiratpur Sahib)	Domestic waste and human ashes
5.	Site-V (Ropar reservoir)	No point source pollution
6.	Site-VI (Ropar)	Wastes from UPM
7.	Site-VII (Ropar)	Wastes from ZPM
8.	Site-VIII (Ludhiana)	No direct source of pollution
9.	Site-IX (Ludhiana)	Industrial wastes from Ludhiana via Buddha-Nallah
10.	Site-X (Jalandhar)	Industrial wastes from Jalandhar via East Bein

PNFC: Punjab Nitrate Fertilizers Corporation; NFL: National Fertilizers Limited; PACL: Punjab Alkalies and Chemicals Limited; UPM: United Paper Mills; ZPM: Zenith Paper Mills.

Collection and Analysis of Sample

The detailed physico-chemical analysis of water sample was carried out according to standard methods of Michael (1984), Trivedi and Goel (1986) and APHA (1989). Based on various physico-chemical parameters, Water Quality Index (WQI) was calculated for each site and water was categorized (Kaur *et al.*, 2001).

Sampling of insects was done from the selected sites of the river by Ekman's dredge. The mud sample was passed through a series of

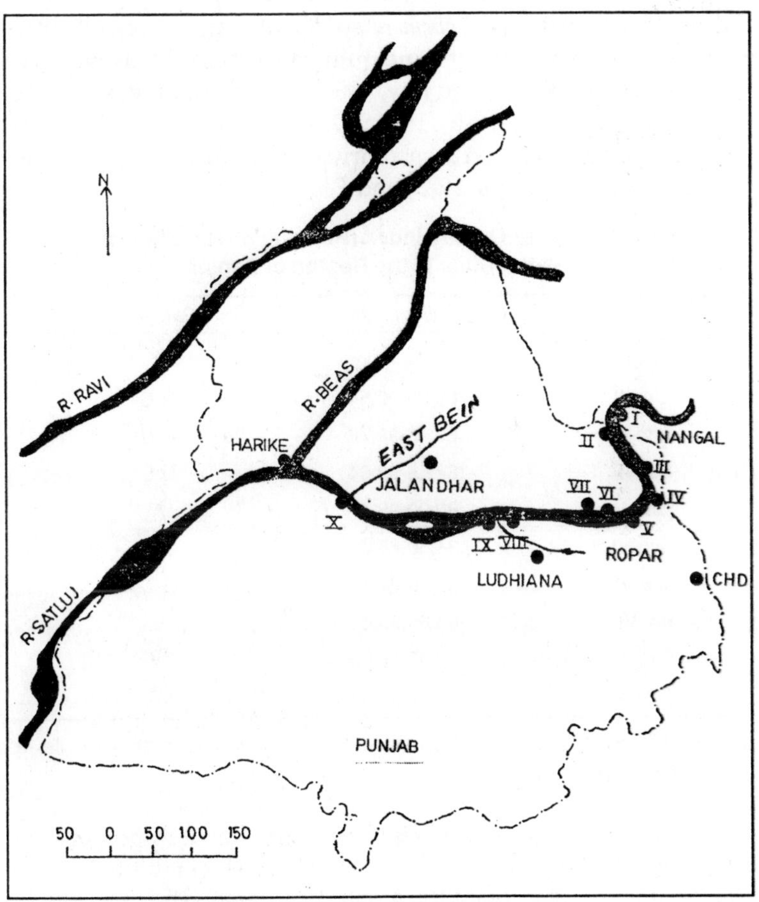

Figure 9.1: Collection Sites on River Satluj

sieves with addition of water on the spot. Animals were picked up with forceps, pippetes and by hand picking, and preserved in 70 per cent alcohal. The identification was done with help of the keys given by Ward and Whipple (1959), Mellanby (1963) and Tonapi (1980).

Results and Discussion

The physico-chemical studies reveal water to be clean at site-I and site-V. Insect diversity is lower at both these sites and is represented by Diptera, Ephemeroptera, Coleoptera, Lepidoptera, Hemiptera and Plecoptera. Species present here are those which are either clean water species such as larvae of *Nymphula* sp., *Helochares lividus* (Das, 1989) and *Ephemera* sp. (Gaufin and Tarzwell, 1956; Verma and Shukla, 1969) or ubiquitous species such as larvae of *Chironomus* sp. (David and Ray, 1966; Das, 1989), Larvae of *Culex* sp. (Badola and Singh, 1981; Das, 1989), *Hydhophilis* sp. (Chowdhary, 1984), *Corixa* sp. (Shrivastava, 1962; David and Ray, 1966; Gaufin, 1974; Das, 1989) and *Perla* sp. (Das, 1989).

Table 9.2: Water Quality Index (Ranges) at Ten Sites of the Satluj River in the Region of Punjab

Sites	*Water Quality Index (Range)*	*Category*
Site- I	84.56-100	A
Site-II	60.03-92.54	B-C
Site-III	67.43-98.78	B-C
Site-IV	67.55-99.04	B-C
Site-V	80.00-99.47	A
Site-VI	34.60-86.78	C-D
Site-VII	34.10-93.12	C-D
Site-VIII	63.65-93.65	B
Site-IX	20.85-66.36	C-E
Site-X	51.97-92.85	B-D

A: Almost Clean; B: Slightly Polluted; C: Moderately Polluted; D: Excessively Polluted; E: Severely Polluted.

At site-II, III and IV, water is slightly to moderately polluted. At site-II fertilizer factory wastes are added which get diluted by the time water reaches at site-III. At site-IV human ashes are added. Among these sites, species diversity is richest at site-III where water is moderately rich in nutrients, slow moving and rich in aquatic plants and thus, is a perfect harbour for the insect growth as is also recorded by Hynes (1970) and Colbo (1985). At site-II lower insect diversity could be due to direct addition of wastes from fertilizer factory whereas at site-IV fast water current and frequent human inference for immersion of human ashes seem to be detrimental for insect fauna. At these sites fauna is represented by clean and ubiquotous species belonging to Ephemeroptera, Hemiptera, Coleoptera, Collembola, Lepidoptera, Plecoptera, Megaloptera, Trichoptera, Hymenoptera and Diptera.

Table 9.3: Distribution of Insect Fauna at Ten Sites of the Satluj River in the Region of Punjab

Sl.No.	*Taxa*	*Sites*									
		I	*II*	*III*	*IV*	*V*	*VI*	*VII*	*VIII*	*IX*	*X*
Order: Diptera											
1.	*Chaoborus* sp. (larvae)		P	P			P			P	
2.	*Chironomus* sp. (larvae)	P	P	P	P	P	P	P	P	P	P
3.	*Culex* sp. (larvae)	P	P	P		P	P	P	P	P	P
4.	*Dicranota* sp. (larvae)									P	
5.	*Eristalis* sp. (larvae)		P	P			P	P		P	P
6.	*Tabanus* sp. (larvae)		P	P							P
7.	*Tanypus* sp. (larvae)			P							
8.	*Tipula* sp. (larvae)		P	P						P	
Order: Coleoptera											
9.	*Dysticus* sp.			P							P
10.	*Helochares lividus* (larvae)		P	P	P	P					
11.	*Hydrophilis* sp.	P	P	P	P					P	
Order: Ephemeroptera											
12.	*Baetis* sp. (larvae)		P	P	P		P	P			
13.	*Ephemera* sp. (larvae)	P		P		P					
Order: Megaloptera											
14.	*Corydalus* sp. (larvae)			P	P						
Order: Hemiptera											
15.	*Corixa* sp. P	P	P							P	

Contd...

Table 9.3–Contd...

Sl.No.	*Taxa*	*Sites*									
		I	*II*	*III*	*IV*	*V*	*VI*	*VII*	*VII*	*IX*	*X*
Order: Trichoptera											
16.	*Hydropsyche* sp. (larvae)			P							P
Order: Collembola											
17	*Isotoma palustris*		P	P	P						P
Order: Lepidoptera											
18	*Nymphula* sp. (larvae)	P	P								
Order: Plecoptera											
19	*Perla* sp. (larvae)		P	P	P	P					
Order: Hymenoptera											
20	*Polynema natans*			P							
	Total no. of taxa	6	13	18	7	5	5	4	2	7	8

P: Present.

At site-VI and VII paper mill effluence is added and water is moderately to excessively polluted. These sites show low insect diversity which is mainly dominated by pollution tolerant Dipterans such as *Chaoborus* sp., *Chironomus* sp. (Jhingran, 1991), *Dicranota* sp., *Eristalis* sp. (David and Ray, 1966; Paramasivam and Sreenivasan, 1981) and *Culex* sp. (Badola and Singh, 1981).

At site-VIII where water is fast flowing and lacks any point source pollution, only two Dipterans, *Chironomus* sp. and *Culex* sp. were found. Site-IX receives effluence from multiple types of industrial units located in Ludhiana via Buddha-Nallah. This is the most polluted site of the Satluj river and remains severely polluted most of the time. Here, insect fauna is dominated by Dipterans which include pollution tolerant species such as *Tipula* sp. (Chowdhary, 1984), *Chaoborus* sp. (Das, 1989), *Chironomus* sp. (Gaufin and Tarzwell, 1956; Husainy, 1965; Mason, 1981; Sinha and Prasad, 1988 and Dhillon *et al.*, 1993a), *Culex* sp. (Badola and Singh, 1981), *Dicranota* sp., and *Eristalis* sp. (David and Ray, 1966; Paramasivam and Sreenivasan, 1981). Their dominance is attributed to the fact that these species possess respiratory adaptability to survive in nutrient rich, oxygen-depleted environment (Gaufin and Tarzwell, 1956; Curry, 1962 and Jhingran, 1991). Site-X is comparatively less polluted and besides Dipterans, a few Hemipteran, Trichoptran, Collembolan and Coleopteran make their appearance.

Distribution pattern of insects in the Satluj river presents a very good example of rivers undergoing rapid trophic evolution as a result of nutrient loading by various pollution sources followed by subsequent self purification along its course. These changes are manifested in change in species composition at various sites. Certain insect species are dominant in some regions of the river and their dominance is dependent on the ecological conditions prevailing there. For example clean water species are found to be dominant at Site-I and V, while pollution tolerant species are found to be dominant at sites-VI, VII, IX and X where intensity of pollution are extreme in terms of loading of pollutants. Species compositions at various sites of the river clearly depict that as the river flows through the Punjab state it gets, progressively polluted.

Acknowledgment

Authors are grateful to PSCST, Chandigarh for financial assistance.

References

A.P.H.A., 1989. *Standard Methods for the Examination of Water and Wastewater*. American Public Health Association, New York.

Badola, S.P. and Singh, H.R., 1981. Hydrobiology of the river Alaknanda of the Garhwal Himalaya. *Indian J. Ecol.*, 8(2) : 269–276.

Chowdhary, S.K., 1984. Studies on bio-ecology of aquatic insects of Sind and Lidder streams of Kashmir, Indian. *J. Ecol.*, 11(1): 160–165.

Colbo, M.H., 1985. Variation in larval black fly populations at three sites in a stream over five years (Diptera: Simuliidae). *Hydrobiologia*, 121: 77–82.

Curry, L.L., 1962. A survey of environmental requirements for the midge (Diptera–Tendipediae). In: *Biological Problems in Water Pollution, Transactions of 3rd Seminar*, (Eds.) Tarzwell, C.M., Usdhew, P.H.S. and Robert, A., Taft Sanitary Engineering Centre, Cincinnati.

Das, S.M., 1989. *Handbook of Limnology and Water Pollution.* South Asian Publishers Pvt. Ltd., New Delhi.

David, A. and Ray, P., 1966. Studies on the pollution of river Daha (N. Bihar) by sugar and distillery wastes. *Environ. Hlth.*, 8: 6–35.

Dhillon, S.S., Kaur, H., Bath, K.S., Mander, G. and Syal, J., 1993a. The Impact of effluence on the zooplankton population of Beas river. *J. Ecotoxicol. Environ. Monit.*, 3(2): 117–120.

Gaufin, A.R., 1974. Biological indices of environmental changes in aquatic habitats. In: *Industrial Pollution,* (Ed.) Sax, I.N. Van Nostrand Reinhold Company, N. York.

Gaufin, A.R. and Tarzwell, C.M., 1956. Aquatic invertebrates as indicators of organic pollution in Lytle Creak. *Sewage and Indust. Wastes*, 28: 908–924.

Husainy, S.U., 1965. Limnological studies of the departmental from Anamalainagar. *Environ. Hlth.*, 7: 24–31.

Hynes, H.B.N., 1970. *The Ecology of Running Water.* Univ. of Toronto Press, Toronto.

Jhingarn, A.G., 1991. Fish in relation to water quality. Paper presented at *Indo-Dutch Meet. on Biomon.* Yard. Dev., New Delhi.

Kaur, H., Syal, J. and Dhillon, S.S., 2001. Water quality index of the river Satluj. *Poll. Res.*, 20(2): 199–204.

Mason, C.F., 1981. *Biology of Freshwater Pollution.* Longman Inc., N. York.

Mellanby, H., 1963. *Animal Life in Fresh Water*, 6th edn. Chapman and Hall Ltd. London.

Michael, P., 1984. *Ecological Methods for Field and Laboratory Investigations.* Tata MacGraw Hill Company Ltd., New Delhi.

Parmasivam, M. and Sreenivasan, A., 1981. Changes in algal flora due to pollution in Cauvery River. *Indian J. Envion Hlth.*, 23(3): 222–238.

Shrivastava, H.N., 1962. Aquatic fauna as indicator of pollution. *Environ. Hlth.*, 4: 106–113.

Sinha, R.K. and Prasad, K., 1988. Ganga basin research project. Buxar–Barh. *Final Tech. Rep. (July, 1985–June, 1988). Patna Univ., Patna.*

Tonapi, G.T., 1980. *Fresh Water Animals of India: An Ecological Approach*. Oxford and IBH Publ. Co., New Delhi.

Trivedy, R.K. and Goel, P.K., 1986. *Chemical and Biological Methods for Water Pollution Studies*. Enviorn. Publs., Karad, India.

Verma, S.R. and Shukla, G.R., 1969. Pollution in a perrenial stream, Khala by the sugar factory effluent near Laksar, Shaharanpur, India. *Enviorn. Hlth.*, 11: 145–162.

Ward, H.B. and Whipple, G.C., 1959. *Fresh Water Biology*, 2nd edn. John Wiley and Sons, Inc., New York, London.

Chapter 10

Insect Fauna of Ropar Headworks Reservoir, Punjab

H. Kaur, K. Sajeev, S. Jasuja and S.S. Dhillon*
Department of Zoology, Punjabi University, Patiala – 147 002

ABSTRACT

Water is the most vital factor on which the living organisms depend. It is basic to the processes that make life possible. Various water bodies, with an attributed name of wetlands, are definable habitats, even though sometimes these arbitrarily change their boundaries. Ropar reservoir, Ramsar Site, came into existence when an irrigation barrage was constructed across the Satluj river at Ropar town during the year 1882 with an objective to build network of canals for irrigation of agricultural land. A complete physico-chemical and biological analysis of water of the Ropar reservoir was carried out. In the present paper, insect fauna of the Ropar reservoir has been discribed and their seasonal abundance and species diversity have been discussed.

Keywords: *Insects, Ropar reservoir, Species diversity, Population density.*

* *Corresponding Author:* E-mail: harbhajankaur@hotmail.com.

Introduction

The Satluj river is one of the main rivers of Punjab. Due to the barraging of the river at Ropar, an area of 800 hectares came under water to form an artificial lake and another 220 hectares were converted into marshes. Ropar reservoir, a Ramsar site, is located at 30°58'–31°02' 03" N latitude and 76°30'–76°33' E longitude at an elevation of 275 meters above mean sea level. The depth of lake varies from 1/2 meters to 6 meters. The Satluj river, Sirsa and numerous other tributaries feed the lake. Ropar reservoir supports a variety of floral and faunal components and is a strategic aquatic ecosystem in the region. The reservoir is also an outlet for the release of hot water generated from the Ropar Thermal Plant which is situated about 10 kms. upstream of the reservoir. In the present paper, distribution of insect fauna in the Ropar reservoir has been described in terms of species diversity and population dynamics.

Materials and Methods

Ropar reservoir harbours a mosaic of habitats which vary in terms of depth, water current, penetration of light, growth of vegetation, human activities, addition of pollutants etc. Keeping in view these ecological features, in all six sites were selected, four in the upstream and two in the downstream of the reservoir. Site-I is located on the right bank of the reservoir and represents the turbulent zone of the reservoir. Water here is unpolluted and is also marked by the absence of aquatic vegetation. Site-II lies downstream of the site-I across the barrage and has stony bed and represents the rapid zone of the reservoir. Here water is shallow, fast flowing, well illuminated and harbours poor aquatic vegetation. Site-III lies approximately in the middle of the upstream portion of the reservoir. Here water is obstructed due to permanent closing of the barrage gates that allows the silt to accumulate and raises the reservoir bed. This region represents the pool zone of the reservoir and is characterized by calm water supporting a rich aquatic vegetation. Site-IV lies on the left bank of the reservoir representing turbulent zone supporting a very poor aquatic vegetation. Site-V lies downstream of the site-IV across the barrage. The bed is stony with pebbles remaining covered with fast flowing water only when barrage gates are open. Site-VI lies about 1 km upstream of the site-IV. This is typical lotic water site showing a definite unidirectional flow and represents rapid zone of the reservoir. The area is occupied

Table 10.1: Physico-chemical Analysis at Six Sites of Ropar Reservoir

Sl.No.	Parameter	Sites					
		I	*II*	*III*	*IV*	*V*	*VI*
1.	Temperature (°C)	18–26	18–28	18–28	18–28	18.5–27.0	18–26
2.	pH	7.1–9.0	7.3–8.8	7.2–9.0	7.2–8.8	7.3–8.9	7.2–8.8
3.	Dissolved Oxygen (mg/L)	4.4–8.4	5.0–8.4	5.2–8.4	5.6–8.2	4.0–8.4	4.8–8.4
4.	Alkalinity (mg/ml)	10–128	20–126	20–140	20–134	20–108	10–124
5.	Hardness (mg/ml)	90–160	80–160	90–160	90–200	110–156	86–180
6.	Chloride (mg/ml)	10–40	8–32	8–30	7–36	7–38	8–46
7.	Fluoride (mg/ml)	0.08–1.10	0.08–1.10	0.06–1.10	0.04–0.96	0.08–1.20	0.01–1.20
8.	Nitrates (mg/L)	0–1.80	0–1.40	0–0.80	0–1.40	0–1.60	0.03–2.10
9.	Nitrite (mg/L)	0–0.42	0–0.42	0–0.56	0–0.38	0–0.29	0–0.28

by dense population of emergent, floating and submerged aquatic plants and is frequently used for cattle wading, washing and bathing purpose.

Water samples were collected from each of the selected sites in 500 ml. plastic containers and physico-chemical parameters were analysed using standard methods given by Michael (1984), Trivedi and Goel (1986) and APHA (1989). For biological analysis, samples were collected from the upstream region of the reservoir with the help of Ekman's dredge. From the shallow downstream sections of the reservoir, insects were collected using foreceps from underneath the stone and pebbles. Hand picking was also used wherever necessary.

Results and Discussion

At the Ropar reservoir both the insect species diversity and population density were found to be moderate throughout the period of study. A total of 27 insect taxa were encountered. These belong to the orders: Plecoptera (nymphs), Hymenoptera (adults), Ephemeroptera (Larvae), Odonata (Nymphs), Hemiptera (adults and nymphs), Coleoptera (adults), Diptera (larvae, pupae and adults), Collembola (adults) and Lepidoptera (Larvae).

From the available data, it is indicated that population density of insects remained comparatively higher during the period extending from March to May and August to October and reached to its maximum in June. During the period of heavy rain (July) the population density declined. During winter period extending from November to February, insects were found to be totally absent. Abundance of insects during summer period was also observed by Munawar (1970), Das (1979), Rai and Datta Munshi (1979), Chowdhary (1984), Sharma (1986) and Kaushik *et al.* (1991). According to Elliot (1967), Macan (1970a) and Hynes (1970) temperature is the most apparent factor affecting the seasonal cycle of insects. Kaur *et al.* (1995) and Bath and Kaur (1998) observed that availability of food and vegetation enhanced the growth of insects during the summer period.

In terms of population density and species diversity of insects at six sites of the Ropar reservoir, following order is depicted:

Site-III > Site-VI > Site-IV > Site-II > Site-V > Site I

Table 10.2: Distribution of Insects at Various Sites of Ropar Reservoir

Sl.No.	*Taxa*	*Sites*					
		I	*II*	*III*	*IV*	*V*	*VI*
Order: Plecoptera							
1.	*Nemoura* sp.	P					P
Order: Hymenoptera							
2.	*Polynema natans*					P	
Order: Ephemeroptera							
3.	*Caenis* sp.			P	P		
4.	*Cleon* sp.			P	P		
5.	*Baetis* sp.			P	P		
Order: Odonata							
6.	*Aeshna* sp.	P		P		P	P
7.	*Anax* sp.	P		P			P
Order: Hemiptera							
8.	*Corixa* sp.	P	P	P		P	P
9.	*Notonecta* sp.			P			P
10.	*Deronectes* sp.		P	P	P	P	
11.	*Perla* sp.			P			
Order: Coleoptera							
12.	*Pelomyxa* sp.	P	P	P	P	P	P
13.	*Psephenus* sp.		P				P
14.	*Dytiscus* sp.		P	P	P		P
15.	*Helochares* sp.				P		
Order: Diptera							
16.	*Chironomus tentans*				P	P	P
17.	*Simulium* sp.			P			
18.	*Alluaudomyia* sp.						P
19.	*Culex* sp.		P	P	P		
20.	*Anopheles* sp.			P			P
21.	*Chaoborus* sp.			P	P		P
22	*Tabanus* sp.		P	P			P
23	*Eristalis* sp.						P
24.	*Palpomyia tibialis*						P
25.	*Stilobezzia* sp.						P
Order: Collembola							
26.	*Isotomurus paliestris*			P			P
Order: Lepidoptera							
27.	*Nymphula* sp.		P				

P: Present.

At site-I insects show lowest diversity as well density as the water flow is swift and lacks vegetation. Site-II and V which lie in the downstream section of reservoir show comparatively lower diversity of insects which are represented by Odonata, Hemiptera, Coleoptera, Hymenoptera and Diptera. This is due to thin aquatic vegetation and frequent changes in water level due to irregular opening and closing of gates of the barrage. At site-IV species diversity is comparatively higher as water gets organically rich by the wastes added from adjcining guest house area. Vegetation however is absent.

At site-III population density and species diversity of insect are highest. The fauna is represented by larvae of *Caenis* sp., *Baetis* sp. and *Cleon* sp. (Ephemeroptera), nymphs of *Aeshna* sp. and *Anax* sp. (Odonata), nymphs of *Notonecta* sp., *Corixa* sp., *Decronectes* sp. and *Perla* sp. (Hemiptera), adults of *Pelomyxa* sp. and *Dytiscus* sp. (Coleoptera), larvae of *Simulium* sp., *Culex* sp. (Diptera) and adults of *Isotomurus palustris* (Collembola). At this site insect fauna is dominated by Odonatans and Dipterans. Here, insect diversity is rich due to the presence of calm, organically rich water and thick vegetation. At site-VI also, population density and species diversity was quite high due to rich organic matter and thick vegetation. The fauna is represented by nymphs of *Nemoura* sp. (Plecoptera), *Aeshna* sp. and *Anax* sp. (Odonata), adults of *Corixa* sp. and *Notonecta* sp. (Hemiptera), *Isotomurus palustris* (Collembola), *Pelomyxa* sp., *Psephenus* sp. and *Dytiscus* sp. (Coleoptera) and larvae of *Chironomus* sp., *Alluaudamyia* sp., *Palpomyia* sp., *Stilobezzia* sp., *Tabanus* sp., *Anopheles* sp., *Chaoborus* sp. and *Eristalis* sp. (Diptera). Aquatic vegetation provides food, shelter and protection as is also observed by Sharma (1986), Kaushik *et al.* (1991), Kaur *et al.* (1995) and Bath and Kaur (1998).

Acknowledgement

Authors are grateful to PSCST, Chandigarh for financial assistance.

References

A.P.H.A., 1989. *Standard Methods for the Examination of Water and Wastewater*. American Public Health Association, New York.

Bath, K.S. and Kaur, H., 1998. Seasonal distribution and population dynamics of aquatic insects in Harike reservoir (Punjab). *J. Ecobiol.*, 10(1): 43–46.

Chowdhary, S.K., 1984. Studies on bio-ecology of aquatic insects in Sind and Lidder streams of Kashmir. *Ind. J. Ecol.*, 11(1): 160–165.

Das, S.M., 1979. Effects of increase land use upon freshwater in the central Himalayas. *Progr. Water Tech.*, 11(6): 195–208.

Elliot, J.M., 1967. The life histories and drifting of the plecoptera and Ephemeroptera in a Dartmoor stream. *J. Anim. Ecol.*, 36: 343–361.

Hynes, H.B.N., 1970. *The Ecology of Running Water.* Toronto University Press, Toronto, pp. 555.

Kaur, H., Dhillon, S.S., Bath, K.S., Kaur, K. and Mander, G., 1995. Invertebrate fauna of freshwater bodies existing in and around Patiala. *J. Env. and Poll.*, 22(4): 163–167.

Kaushik, B., Sharma, S. and Saxena, D.N., 1991. Ecological studies of certain polluted lentic waters of Gwalior region with reference to aquatic insect communities. In : *Current Trends in Limnology–* I, (Ed.) N.K. Shastree, pp. 185–200.

Macan, T.T., 1970a. A key of the nymphs of British species of Ephemeroptera with notes on their ecology. *Sci. Pub. Fresh. Wat. Biol. Asso.*, No. 20.

Michael, P., 1984. *Ecological Methods for Field and Laboratory Investigations.* Tata MacGraw Hill Company Ltd., New Delhi.

Munwar, M., 1970. Limnological studies on fresh water ponds of Hyderabad. I Biotope. *Hydrobiologia*, 35: 127–162.

Rai, D.N. and Datta Munshi, J. S., 1979. The influence of thick floating vegetation (Eichhornia cressioes) on the physico-chemical environment of a fresh water wetland. *Hydrobiologia*, 65: 65–69.

Sharma, R.C., 1986. Effect of physico-chemical factors on benthic fauna of Bhagirathi river, Garhwal Himalayas. *Ind.J. Ecol.*, 13(1): 133–137.

Trivedy, R.K. and Goel, P.K., 1986. *Chemical and Biological Methods for Water Pollution Studies.* Environ. Publs., Karad, India.

Chapter 11

Insect and Aquatic Environment

S. Subhashini, K. Logan Kumar and S. Logaswamy

PG and Research Department of Zoology,

Kongunadu Arts and Science College, Coimbatore

Introduction

Insects are invertebrates (animals without a backbone) that are part of the larger group of animals called arthropods. Arthropod means "joint footed." That name was given to these animals because all of the arthropods have legs with joints that are something like our elbows and knees. Some other arthropod relatives of insects are crayfish, crabs, lobsters, millipedes, centipedes, scorpions, spiders, and ticks. Most insects are terrestrial (live on land), and are found in places such as trees, shrubs, flowers, rocks, logs, soil, buildings, and especially our gardens. Everyone is familiar with common terrestrial insects such as butterflies, moths, beetles, ants, bees, wasps, grasshoppers, crickets, cockroaches, and flies. There are also many kinds of insects that live in the water. These are called aquatic insects, and they are often not seen unless you explore places such as puddles, ponds, lakes, ditches, streams, and lakes. There are many different kinds of aquatic insects and almost every type of freshwater environment will have some kind of aquatic insect living in it. The diversity of insects can only be described as amazing. More than half of all known species of living things (microbes, plants, and animals) are insects.

There are about 751,000 known species of insects, which is about three-fourths of all species of animals on the planet. While most insects live on land, their diversity also includes many species that are aquatic. Just about anywhere you go on the planet, there is some kind of insect that will live in almost any place that stays wet for a week or so. Aquatic insects are important food for fish and waterfowl. They also play important roles in keeping freshwater ecosystems functioning properly. Many species of aquatic insects are very susceptible to pollution or alteration of their habitat. In fact, aquatic insects are the group of living things used most commonly for monitoring the health of aquatic environments.

Habit and Habitat

One of the most amazing things about aquatic insects is the diversity of habitats where they live. There is no body of water that is too small, too large, too cold, too hot, too muddy, with oxygen too low, with currents too fast, or even with too much pollution for some kind of aquatic insect to live there. About the only restriction to where they live is that they do not usually inhabit the salty water of marine environments, such as oceans and bays. However, there are even a couple of unusual aquatic insects that live on coral reefs and in tide pools of marine environments. Estuaries, where the fresh water of rivers mixes with the salt water of oceans, are home to quite a few kinds of aquatic insects. Anyone who has been to the beach knows about the kinds of mosquitoes that breed in the salt marshes near the beach. Not all kinds of aquatic insects live in all types of freshwater habitats.

The most favourable habitats, and the ones where you can collect the most kinds of aquatic insects, are the edges of ponds and lakes and the sections of streams and rivers where the water is flowing fast enough to splash (riffles). In both standing and flowing freshwater habitats, the most different kinds of aquatic insects will be found in water that is less than three feet deep and can be easily waded. Aquatic insects have a variety of special adaptations for moving around or staying in one place within their habitat. Some are agile swimmers by means of streamlined bodies with long legs or tails, while others climb around on aquatic plants by means of long thin bodies. Some sprawl on top of soft mucky bottom without sinking in because their bodies are flat and their legs extend out from the sides. Others are able to burrow down into soft mucky bottom because they have special structures on their bodies, such as legs that look like shovels or points projecting in front of their heads.

Still others can cling to rocks and logs in very swift current because their bodies are very flat and the current just passes over them without knocking them off. Other clingers stay put by using special suckers or by gluing themselves down with sticky silk that they produce. Lastly, many aquatic insects like to crawl around in the tiny spaces among rocks, sticks, and dead leaves. Because aquatic insects are small and highly specialized, different kinds are often found in small areas with similar features, which are called microhabitats. Examples of microhabitats where you will probably find different aquatic insects are: cobble rocks (about the size of your fist or head), gravel, sand, muck, accumulations of dead leaves and twigs, live plants, and grasses and tree roots that extend into the water from land. Different microhabitats, with different aquatic insects living in them, occur very close together, perhaps within one step of each other. Aquatic insects even live in temporary habitats, such as small streams or ponds that dry up in the summer. If they are adults, they can simply fly to another place with water. Some immature aquatic insects that cannot yet fly will burrow down into the bottom where it is damp and go into an inactive state, something like animals hibernating over winter.

However, most aquatic insects that live in temporary habitats are "programmed" to stay in their eggs, where they are protected, until the time of year when water is present. Different insect species eat different types of food. Herbivores eat nothing but plants. Algae, the "slippery green stuff" that grows on rocks at the bottom of streams, is a favourite food of many herbivores. Predators eat other animals. Predatory aquatic insects eat other insects, fish and even small frogs. Omnivores feed on plants and animals. Scavengers eat dead plant and animal matter. The foods of aquatic insects are just as diverse as the habitats they live in. Although individual kinds of aquatic insects may only eat one type of food, all organic material in the water, living and dead, is eaten by some kind of aquatic insect.

Scientists have found it informative to categorize aquatic insects according to how they obtain their food for studying the ecology of freshwater ecosystems. These categories are called functional feeding groups. Scrapers have special mouthparts that remove algae growing on the surface of rocks or other solid objects. These mouthparts work like a sharp blade to remove the outermost layer of algae, which is attached very tightly but is very nutritious for those insects equipped to remove it. Collectors acquire small pieces of decaying plant material (detritus). Some kinds use long hairs on their head or legs or silk nets to filter these small particles out of the water. Other kinds of collectors use their mouthparts to gather fine particles lying on the bottom and shove this material into their mouths. Shredders have mouthparts that are designed to nibble off pieces of soft vegetation, such as leaves, flowers, or twigs, and grind up this material.

Most aquatic insects shred pieces of vegetation that have dropped off of plants and are decaying. Most of this material comes from trees and shrubs that grow on land at the edge of the water. Only a few kinds of aquatic insects feed on parts of live plants that grow under the water. Predators feed on other animals that are alive. Predators often have special structures for catching and subduing their prey, such as strong jaws with teeth, a sharp beak, or spiny legs. Predators eat other invertebrates most of the time, but some are large and strong enough to catch small vertebrates, such as fish and tadpoles.

External Features and Adaptations

Like the other arthropod relatives, insects have their skeleton on the outside of their body (exoskeleton). This outside skeleton is thick, and often hard, so that it can protect the insect, much like our skin protects us. However, an insect's exoskeleton does not grow along with the insect. As an insect grows it must shed its skin and grow a new one in order to get larger. There are several features of insects' bodies that make them different from the other arthropods. The body of an insect is made up of three sections. The head is at the front end of the body. The thorax is in the middle of the body and is usually larger than the head. The abdomen is the rear section of the body and is usually as long, or longer, than the head and thorax together. You can usually see individual segments (up to 10) in the abdomen, but the individual segments in the head and thorax are usually fused together and cannot be easily distinguished.

The head contains structures for eating and sensing the world that an insect lives in. Insects have several different mouthparts that are specialized for tasting, obtaining, and breaking up food. There are two antennae (feelers) on the head, one on each side. There are usually two large compound eyes, which contain thousands of small individual eye cells. The thorax contains the structures that insects use to move around. There are six legs, arranged with three on each side of the body. Some of the legs may be constructed for special movements or purposes, such as running, jumping, digging, or catching food. Most insects have four wings on the thorax. Insects are the only arthropods that can fly, which is the main reason they are more widespread and more diverse than any of the other arthropods. The abdomen does not have many structures on the outside, except for some structures at the end that are specialized for mating and laying eggs.

Many insects have two tails on the rear of the abdomen, which they often use to sense vibrations. The distinguishing features described above apply to adult insects. For many aquatic insects, only the immature stages (babies) live in the water. Sometimes the immature stages do not have the same features as adult insects. Immature insects never have working wings, and some may not have compound eyes, jointed legs, or a distinguishable thorax section. Some immature insects look more like worms than insects. However, wormlike immature insects will always have an obvious head, or at least some noticeable structures sticking out from the head, such as mouthparts or antennae, while worms never have a head or head structures. Aquatic insects move around in the water in a variety of ways. Some simply crawl along the rocks and mud on the stream bottom. Dragonfly nymphs move around by forcing water out through their abdomen, just like a jet engine. Water boatmen use their long hairy legs to paddle through the water, much like a rowboat. The waxy surface of the water strider's legs allow them to "skate" across the surface of the water. Insects lacking the ability to move themselves through the water simply drift with the current.

Insects that live on land breathe air through holes in their bodies called spiracles. This would not work very well for insects that live in the water, so aquatic insects have special adaptations for breathing in the water without drowning. The most common way for aquatic insects to breathe effectively underwater is to use oxygen that is dissolved in the water rather than oxygen that is a gas in the air. Many aquatic insects, especially during their immature stages, have gills similar to fish for obtaining dissolved oxygen. The gills of aquatic insects are located on the outside of their body in various locations. Gills come in various shapes, but many are flat oval plates or tufts of small filaments.

Other aquatic insects have a soft flexible external skeleton that simply allows dissolved oxygen to pass from the water into their body all over their body surface. Some kinds still use the holes in their bodies to get oxygen from the air. They just keep the holes shut while they are underwater, and only open them when they come to the surface to take in a breath of air. Some kinds take a bubble of air underwater and breathe out of the bubble, which allows them to stay underwater longer.

Insects have reproduced their remarkable adaptability so easily seen on land within the aquatic environment. Certain insect orders have immature stages that are extensively aquatic, such orders are the Ephemeroptera, Plecoptera, Odonata and the Trichoptera. Many other orders also include species that have incorporated an aquatic developmental stage into their life cycle. The aquatic environment creates very different restraints from that of a terrestrial one. Respiration and osmoregulation present very different problems to aquatic insects compared to terrestrial insects. The prominent physical difference between water and air is the drop in available oxygen in water, the increase in density of medium, and the severe currents that can occur in lotic (flowing) waters.

Tracheated cuticular outgrowths are a common morphological adaptation seen in many hydroneustic (living under the surface) insects. These bodily extensions augment the tracheal system, thus giving an increased exchange surface. According to species, gills can be situated abdominally, caudally, at the base of the legs, the neck, mentum, maxillae, the anus, and the rectum, as seen in Anisoptera (dragonfly and damselfly). This closed tracheal system with its non-functioning spiracles allows oxygen to diffuse across the very thin cuticle of the gill area. From this region it can quickly diffuse throughout the body tissues. The incompressible tracheal system prevents collapse as oxygen is used and external pressure increases.

The vast majority of insects that obtain oxygen from solution in the surrounding water retain the tracheal system seen in terrestrial insects. This is because oxygen can be drawn out of solution and into the gaseous state. This allows a much faster diffusion rate than if it was in solution within the insects haemolymph.

Many insects have managed to exist within an aquatic environment without the need to obtain oxygen from the water surrounding them. These aeropneustic insects use respiratory siphons to gain direct connection to a source of air. This may be through the water surface as seen in larval Culicidae or by tapping into specific aquatic plant arenchyma tissue, as seen in the mosquito larvae of *Mansonia,* and other Diptera larvae such as *Chysogaster* and *Notiphila*. The main physical difficulty with obtaining oxygen from the surface is the surface tension of the water, and prevention of water entry into the spiracles.

Insects using this respiratory mechanism have developed hydrofuge surfaces which lower the surface tension and prevent water entering into the spiracles. Oily secretions are produced by perispiracular glands in some dipteran larvae. In *Notonecta* there is an arrangement of non-wettable hairs which close over the spiracle when submerged and open like a fan at the water surface, spreading the surface tension and allowing oxygen into the tracheal system. The main advantage of obtaining oxygen independently from the water is that the insect is unaffected by the waters oxygen tension. This allows virtually any water pool to become available for larval development. This is a major factor in the success of mosquitoes and one of the main obstacles in implementing effective control.

Due to the internal chemical composition of aquatic insects osmoregulation is the converse of that seen in terrestrial insects. Being hypertonic to their environment water passes through the cuticle and into the insects. One method of removing incoming excess water is the ample production of urine, in many cases this is also combined with amounts of ammonia. Salt reabsorbtion also occurs in the gut and Na and Cl ions are known to be reabsorbed by an active uptake process within the rectum. Certain larvae are able to uptake salts from very dilute solutions using active processes. This can be seen in *Aedes, Culex* and *Chironomus* which use anal papillae as the uptake site.

Larval uptake of salts from such dilute solutions maybe due to their increased loss through their permeable cuticle. Many larvae lack the lipid layer seen in the adults. Chloride epithelia are used for ion uptake in Odonata and Diptera. These are localised areas of integument which actively take up and replace ions from the surrounding water. Aquatic insects of marine habitats have to deal with extreme osmotic pressures.

The neustron (water surface) can also provide a habitat for certain species, such as pond skaters. Oxygen is obtained by aquatic insects either by taking that available in abundance above the water, or by uptake from oxygen in solution in the surrounding water. In some cases this has resulted in the production of respiratory pigment, seen in chironomid larvae. Neuston insects live on the surface tension of water. Neuston insects therefore tend to be lightly built with splayed limbs to spread their weight. Features seen on gerrids to enable movement on the surface film include tarsal segments which are flattened and covered in hydrofuge setae to prevent wetting; a body held well clear of the water surface; and raised claws which prevents piercing of the surface film.

The cuticle of many Dytiscidae water beetles is considerably hydrophobic. This results in a high angle between the beetle and the water surface when the beetle is at rest. This has both advantages and disadvantages for the beetle. It makes swimming downwards difficult, but returning to the surface to replenish a physical gill becomes very easy, as does the ability to escape the surface tension in order to disperse. Water current has a direct effect on aquatic insects. This is seen essentially in those insects that inhabit lotic waters. Due mainly to their small size and light weight a planktonic or nektonic existence is not feasible in strong currents. This leaves the benthic environment, and many insects show various morphological adaptations suiting them to this environment.

Dorso-ventral flattening is seen in many aquatic insects, *Ephemerella doddsi* is one example of this. This body shape allows the insect to exploit the boundary layer effect. This is the term for the thin layer of water directly above the substrate, varying from 1 to 3mm in height. In this region the current becomes greatly reduced and therefore the drag exerted on the insect within this zone is considerably less than one outside it. This allows safer movement over the substrate and less energy expenditure to remain stationary. Dorso-ventral flattening allows an insect to also move between the substrate, either in search of food or to escape predation or spate waters. Other insects have opted for a fusiform shape which is anchored by the legs and orientated into the current thus offering the least resistance.

Baetid mayflies are good examples of this. *Ephemerella doddsi* have an anchoring device on their ventral surface formed by an outer ring of setae filled with backward pointing setae. This structure allows the insect to maintain a purchase on the substrate in strong currents. Insects of the order Ephemerotera, Diptera, and Trichoptera have used the current and its associated debris to provide food material. They achieve this by various filtration methods, either by producing silken nets, seen in many caddis fly larvae, or from extensive setae on the legs or mouth parts which act as nets. Food particles are combed from the setae, a method used by the *Isonychia* genus. Nets are frequently eaten and so caught particle are ingested.

Egg laying in aquatic insects shows two main divisions, those that are submerged and lie either on the substrate or fixed to plants, and those that remain in contact with the surface. Some Anisoptera eggs sink to the bottom and remain there until they hatch. Others lay their eggs on over hanging plants so that when they hatch the larvae drop into the water. Certain species of Zygoptera submerge themselves in order to oviposit their eggs under water attached to emergent plants. Many mosquito species have eggs that have floats which keeps them on the water surface. *Culex* lay eggs in batches of up to 300 packed together as a raft. The eggs stand upright with the hydrophilic micropyle cup on the water surface and the hydrophobic chorion exposed to the atmosphere. The downward facing micropyle allows the larvae to hatch and enter straight into the water. These surface eggs mean that the oxygen tension of the water does not effect development of the eggs, a case which is also seen to so extent in the larvae.

Camouflage is an adaptation that provides protection from predators. The body colour and pattern of many aquatic insects allows them to blend in with their surroundings. If a predator can't see them, it won't eat them! Camouflage is an important adaptation for many stream creatures, including salmon and trout. Atlantic salmon even have the ability to change colour to match their background.

Reproduction and Development

Only adult insects are capable of reproducing, and most aquatic insects spend their adult stage out of the water. After mating on land, females return to the water to deposity her eggs. Eggs are usually stuck on solid objects under water, but a few kinds deposit the eggs on trees or rocks above the water. Eggs usually hatch within a few days or weeks, but some may be programmed to not hatch for many months. A delay in hatching allows an aquatic insect to live in habitats that are too hot, too cold, or dry up during part of the year.

Different kinds of aquatic insects require anywhere from a few weeks to several years to develop into adults. Most grow and develop steadily, but some may go into an inactive state to endure harsh environmental conditions. It is most common for aquatic insects to produce one generation per year, with the adults emerging during the warm months. Some of the larger kinds, or those living in cold environments, require two to three years to develop from egg to adult. Some kinds with short developmental times may produce more than one generation per year.

In water you will come across many insects, often perfectly adapted to the aquatic environment. Some species are entirely aquatic, whereas other insects only live in the water during their larval stages or as nymphs. When insects undergo a metamorphosis we call the immature form larva. When they gradually transform via moults into the adult form the young stages are called nymphs. The life of an aquatic insect begins when the parent lays eggs in the water. Immature aquatic insects become adults after undergoing metamorphosis, a process which can radically change the insect's appearance.

Immature aquatic insects are called either larvae or nymphs. A nymph looks somewhat like the adult stage of the insect. Larvae look very different from the adult. Following metamorphosis many species of adult insects remain in the water, spending their entire lives there. Many others leave the water environment and become land based. Mosquitos, dragonflies and black flies are three common insects that are born underwater. As with all forms of life in a stream, aquatic insects have developed certain unique adaptations that allow them to breathe, move, find food and otherwise survive.

Insects, like all arthropods, must shed their protective external skeleton periodically in order to get larger. Before they shed their old skin, they grow a new one underneath. The new larger skin is soft and folded, so that it can fit inside the old smaller skin. They also absorb much of the old skin and recycle the materials it contained. To shed the old skin, aquatic insects gulp water to make themselves larger, and they push from the inside on the top of the thorax. The old exoskeleton is thinner there, so it splits and allows the insect to climb out. After getting out of the old skin, the insect continues to gulp water to make itself larger by stretching out the wrinkles in the new soft skin. After a few hours, or in some cases a few days, the new skin hardens into another protective exoskeleton.

Different kinds of aquatic insects shed their skin anywhere from three to 45 times. This is a dangerous time in the life of an aquatic insect, and many of them die while they are waiting for their new skin to harden. The new soft exoskeleton is subject to damage, and it does not allow the insect to move and hide very well. As insects grow from eggs into adults, they go through a series of developmental changes that are called metamorphosis. There are two basic types of metamorphosis in aquatic insects. It is important to know about these two types because metamorphosis is used to help identify immature insects and it explains a lot of the biological activities they engage in during their lives. One type of metamorphosis is incomplete (also called hemimetabolous).

Aquatic insects with incomplete metamorphosis emerge from the egg looking a lot like miniature versions of the adults, minus wings. They have compound eyes on the head and jointed legs on the thorax. The wings develop in projections on the thorax (wing pads) and get a little larger each time the insect sheds its skin. Scientists have begun to use the term larvae for the immature stages of aquatic insects with incomplete metamorphosis, but for many years they were called nymphs or naiads. The other type of metamorphosis is complete (also called holometabolous). Immature aquatic insects with complete metamorphosis do not look anything like the adults they turn into.

At most they have only a simple eyespot, or a small cluster of eyespots, on the head, and some have no eyes at all. They may or may not have jointed legs on the thorax. The wings develop inside the body, so no wing pads are visible until just before the insect becomes an adult, insects with complete metamorphosis go into an inactive stage before they become adults. During the inactive stage is when compound eyes, jointed legs, and wing pads first show up. The active immature stages are called larvae, and the inactive stage is called a pupa. Some aquatic insects crawl out of the water for the pupa stage, while others spend this stage in the water.

Problems Faced by Aquatic Insects

Hardly any species of aquatic insects have been listed as endangered or threatened. However, the reason for this is that studying the distribution and population numbers of such a diverse group of organisms is an overwhelming task. Also, most people do not understand and appreciate the importance of these small creatures in freshwater ecosystems. It is certain that many species of aquatic insects are threatened and perhaps on the verge of extinction. These are most likely to be species that have narrow ecological requirements, and, thus, live in unique habitats that have not been disturbed by human activities.

In the past, aquatic insects were severely reduced in many bodies of water by discharges of toxic substances, such as those from manufacturing plants and mines. Also, overloaded sewage treatment plants discharged human waste, which used up all of the dissolved oxygen when it decayed. However, aquatic insects still face a great threat from nonpoint source pollution. This widespread problem comes mainly from excessive amounts of nutrients and sediment that get into streams, rivers, ponds, and lakes from so many different sources that it is hard to track them all down. The human activities that are responsible for most of the nonpoint source pollution are agriculture, forestry, and urban development.

Many kinds of aquatic insects are eliminated because the excess nutrients and sediment cover the surfaces where aquatic insects need to hold on or clog the small spaces where they need to hide. In addition, these pollutants cause the water to have less dissolved oxygen. Other current nonpoint source problems for aquatic insects include warm water temperature caused by removing the trees that grow along streams and the introduction of toxic substances that wash off of city streets and people's lawns.

Biomonitoring

Aquatic insects can reveal a great deal about the health of a stream, pond, river or lake. Many species are very sensitive to water quality. They are particularly sensitive to the amount of oxygen in the water. A healthy stream generally has a large variety of aquatic insect species. Only the hardiest of insects can survive in a polluted aquatic environment. Water quality monitoring is a popular environmental conservation activity. One aspect of testing for water quality generally involves determining the variety and number of aquatic insects and other invertebrates in a body of water.

If you have been wondering about the health of a nearby water body you might begin to get some answers by doing a survey of the insects that live there. Aquatic insects are probably best known for their ability to tell us about the water quality in a particular environment. Some of them are very sensitive to pollution, while others are tolerant. If you take a sample of the aquatic insects in a particular place, and analyze the sample in terms of the sensitive kinds versus tolerant kinds, you can get a good measure of the environmental health. Healthy aquatic environments have a lot of different sensitive kinds, while polluted environments have only a few kinds of tolerant aquatic insects. This process is called biological monitoring (or biomonitoring) and is commonly done by government agencies as well as citizen volunteer organizations.

However, the use of aquatic insects for biomonitoring is not the main reason that they are important. Because there are so many different kinds of aquatic insects and their activities are so diverse, they perform essential roles that keep freshwater ecosystems functioning properly. A good analogy would be the various jobs of factory workers on an assembly line that are necessary to make sure that a manufacturing plant turns out plenty of good products. Some of the aquatic insects are responsible for breaking down the dead leaves and other plant parts that fall into bodies of water from land.

This material provides the base of the food chain in some aquatic environments, especially small streams in forests. Some scrape the algae that grow on all firm surfaces in water, such as rocks, logs, and the leaves and stems of live rooted plants. This layer of algae, which produces much oxygen and food for other organisms, is more productive if it is kept thin by the grazing of aquatic insects and other invertebrates. Other kinds of aquatic insects are specialized for filtering fine particles that are suspended in the water. This is useful because it helps to keep the water clear enough for light to penetrate where algae and other plants are growing on the bottom. Other kinds mix the soft bottom sediments as they burrow in search of food.

This makes the bottom healthier for organisms because it puts oxygen from the water into the bottom. Lastly, the aquatic insects that are predators reduce the numbers of other invertebrates and help keep a balance among the different kinds of organisms and the food that is available. Even if aquatic insects did not perform these important jobs in aquatic ecosystems, they would still be useful just because collecting and observing them is so easy and so much fun. Aquatic insects are an excellent way to get people of all ages interested in nature and conservation of natural resources.

Collection of Aquatic Insects

Aquatic insects are always easy to collect because they are so numerous and they live in so many different places. The simplest way is just to pick up objects in the water, such as rocks, plants, sticks, or leaves, and examine the material. Sometimes it works best to place the object in a shallow pan with clean water. You can see aquatic insects with your naked eye, but a magnifying glass might help to find some of the smaller ones. You will probably collect more organisms by using a net with fairly fine mesh. These are available in stores or you can build your own from materials around the house. If there is current, hold the net in the water and move the habitat where you think aquatic insects might live upstream from the net.

It works best if you move the habitat with your hands, but it will also work to kick with your feet (it just damages more of the organisms). Either way, aquatic insects and some of the material where they were living will wash right into your net. If you are in standing water, you will need to move the net in the places where the aquatic insects live. Usually short pokes with the network best. After collecting material in the net, empty the contents into a shallow pan with clean water to find the aquatic insects.

There is no reason to kill aquatic insects unnecessarily, so it is always best to release them back into the water alive after studying them for a while. However, since most aquatic insects are very abundant, it is acceptable to make a preserved collection of common ones for educational purposes. The best way to preserve aquatic insects is to place them in alcohol in a small clear container with a tight fitting lid. Rubbing alcohol (isopropanol) works fine. You should put a paper label in the container stating what kind of aquatic insect it is, as well as where and when you collected the organism.

Chapter 12

Statistical Approach to Monthly Variations of Physico-chemical Factors at Lower Lake of Bhopal in Relation to Insect Fauna

Madhulika Singh[1] *and K. Borana*[2]

[1]*Sadhu Vaswani College, Bairagarh, Bhopal*

[2]*Department of Applied Aquaculture, Barkatullah University, Bhopal*

ABSTRACT

A study on monthly variations of insects was made in Lower lake of Bhopal, during June 2002 to May 2004. Coleoptera, Diptera, Ephemeroptera, Hemiptera, Odonata and Trichoptera dominated insects. The density of total population of insects ranged between 201 org/sqm to 450 org/sqm having higher percentage composition of Coleoptera during most of the study period. Physico-chemical factors (water temperature, conductivity, pH and dissolved oxygen) were also estimated and their correlation with total insect density was established.

Introduction

Aquatic insects play important role in the aquatic community. They are involved in the mineralisation and recycling of organic matter. Bioindicators are organisms employed to assess general toxicity of the environment. These organisms by their presence or absence and in all features of their phenotype and physiology serve as an indicator of environmental status. Thus both the positive and negative facts of the environment can be monitored through biological system. Several biotic and abiotic factors influence the productivity of aquatic insects and an understanding of these factors has become an essential pre-requisite for maintaining not only the aquatic insects but also the animals that depend on them for their food.

The Indian aquatic insects are described by several authors some important contributions are by Tonapi and Rao (1970), Reddy and Pandian (1972), Kapoor (1974), Kumar (1980), Mishra (1985), Oomachan (1988), Saxena *et al.* (1989), Kumar (1993), Sridharan *et al.* (2000) and Kandibane *et al.* (2005).

Materials and Methods

The study of the physico-chemical parameters and insects were carried out in Lower lake of Bhopal from June 2002 to May 2004. The mud samples were collected with Ekman Dredge sampler at monthly intervals. After that mud was sieved out through 0.5 mm mesh size sieves according to Jonasson (1955) and Hovgaurd (1963) method. Physico-chemical parameters such as conductivity, pH and dissolved oxygen were estimated according to APHA (1995). Correlation between insects and certain physico-chemical parameters were computed.

Results

The insects were represented by six orders *viz.*, Coleoptera, Diptera, Ephemeroptera, Hemiptera, Odonata and Trichoptera (Table 12.1). Altogether 38 insects species were identified in which 12 belong to Coleoptera, 5 belong to Diptera, 4 belong to Ephemeroptera, 8 each to Hemiptera and Odonata and 1 belong to Trichoptera (Table 12.2).

The total insect density ranged from 202 org/sqm in September 2003 to 450 org/sqm in April 2004 of all the studied insect population in this lake, Coleoptera was the most dominant and abundant order and it ranged between 55 org/sqm in December 2002 to 237 org/sqm in May 2004. Dipterans ranged from 2 org/sqm in July 2002 to 64 org/sqm in October 2002, Ephemeropterans ranged from 10 org/sqm in September 2003 to 102 org/sqm in April 2003, Hemipterans

Table 12.1: Physico-chemical Factors and Monthly Density on Insects Order from June 2002 to May 2004

Month	*Water Temp. °C*	*Conduct. µmhos/cm*	*pH*	*D.O. ppm*	*Total Insect org/m²*	*Coleop. org/m²*	*Dipt. org/m²*	*Ephem. org/m²*	*Hemip. org/m²*	*Odona org/m²*	*Trichop. org/m²*
Jun. 02	29.5	510	7.8	6.4	268	169	5	30	25	32	7
Jul.	24.6	440	7.4	5.2	207	130	2	20	35	17	3
Aug.	25.5	490	7.5	6.8	219	85	25	32	47	22	8
Sep.	25.0	460	8.0	6.4	280	77	95	48	77	28	5
Oct.	25.2	470	8.7	6.0	315	66	64	53	93	35	4
Nov.	24.2	500	8.2	6.4	398	67	56	61	95	56	6
Dec.	21.0	490	7.9	7.2	314	55	52	58	82	57	10
Jan.03	21.5	500	7.5	7.6	290	65	40	49	53	71	12
Feb.	22.0	530	7.6	6.8	307	96	23	68	32	74	14
Mar.	23.0	580	7.5	6.8	322	134	4	79	26	72	17
Apr.	27.0	639	8.0	6.4	411	191	14	102	19	77	8
May.	28.0	658	8.4	6.0	391	219	7	82	24	55	4
Jun.	29.3	610	8.8	6.6	324	218	12	49	16	26	3
Jul.	28.5	544	8.6	6.0	267	179	19	21	21	23	4
Aug.	27.3	570	8.4	6.4	227	109	20	17	57	22	2

Contd...

Table 12.1–Contd...

Month	*Water Temp. °C*	*Conduct. µmhos/cm*	*pH*	*D.O. ppm*	*Total Insect org/m²*	*Coleop. org/m²*	*Dipt. org/m²*	*Ephem. org/m²*	*Hemip. org/m²*	*Odona org/m²*	*Trichop. org/m²*
Sep.	27.0	556	7.6	6.8	202	83	21	10	61	23	3
Oct.	27.5	548	8.1	6.0	263	72	52	29	84	20	6
Nov.	25.0	528	7.8	6.4	290	71	45	39	81	43	11
Dec.	23.0	511	7.6	6.8	273	85	37	37	52	53	9
Jan.04	20.0	501	7.7	7.2	272	103	34	36	50	46	3
Feb.	22.0	533	8.6	7.6	273	109	22	33	47	64	8
Mar.	25.0	608	9.2	7.2	350	155	26	48	47	68	6
Apr.	28.0	632	4.7	6.4	450	219	19	79	51	75	5
May.	29.5	652	9.6	6.8	417	237	14	65	44	49	8

ranged from 16 org/sqm in June 2003 to 95 org/sqm in November 2002, Odonatans ranged from 17 org/sqm in July 2002 to 77 org/sqm in April 2003. The last dominant group of insect was Trichoptera and it ranged between 2 org/sqm in August 2003 to 17 org/sqm in March 2003.

Table 12.2: List of the Insects of Different Orders Encountered Throughout the Year

Sl.No.	*Order*	*Genera*
1.	Coleoptera	*Berosus, Cybisterr, Dineutus ind icus, Dineutus unidentatus, Haliplus, Hydaticus, Hydrobius, Hydrophilus, Hydrovatus, Laccodytes, Laccophilus and Sternoplus*
2.	Diptera	*Eristalis, Psychoda, Stratiomys, Tendipes tendipes and Tendipes kifferulus*
3.	Ephemeroptera	*Ameletus, Baetis, Caenis and Cloeon*
4.	Hemiptera	*Belostoma, Corixa minor, Corixa varicunda Diplonynchus, Gerris, Micronecta minthe, Ranatra lenearis and Ranatra varipes*
5.	Odonata	*Boyeria, Brachythemis, Coenagrion, Enollagma, Hagenius, Hesperagrion, Libellula and Trithemis*
6.	Trichoptera	*Hydropsyche*

The correlation coefficient values of total insects with certain physico-chemjcal parameters were established. Total insects showed direct and significant correlation with conductivity ($r = 0.68, p << 1$), pH ($r = 0.64, P << 1$)and dissolved oxygen ($r = 0.72, P << 1$).

Discussion

The present study showed that the highest insect population was observed in suromer months (April and May) in Lower Lake. Michael (1968) recorded the presence of *Chironomus* larvae from October to May. Greenfield and Ireland (1978) and Stoneburner and Smock (1979) recorded the highest number of insects in May. Nebeker (1971) reported that increased temperature accelerate the emergence of aquatic insects.

The majority of insect species found in average eutrophic water body can tolerate a broad range of physico-chemical conditions of water, which render them useless and some of them adopt and become indicators of the water quality (Roback, 1962). David and Ray (1966) observed that Mayflies and Damselfly Nymphs and Hemipteran bugs were found inhabiting in the comparatively well-oxygenated zone. Crosswill (1949) observed that Ephemeropteran nymphs and larvae, Odonates, Dipterans beetles and water bugs require low oxygen while some other caddishfly and Mayfly larvae require higher oxygen contents. According to Roback (1962) the Odonata are not sensitive group, they cannot tolerate high changes in pH.

It has been well established that insect communities depend directly or indirectly on different physical and chemical factors, which interact with each other. In the present study total insect density showed direct and significant correlation with conductivity, pH and dissolved oxygen.

References

American public Health Association (APHA), 1995. *Standard Methods for the Examination of Water and Wastewater*, 18th Edn. American Public Health Association, Washington, D.C.

Crosswell, H., 1949. Value of the bottom samples in demonstrating the effects of pollution on fish food organisms and fish in Shenandoan river. *The Prog. Fish. Cult.* 11: 217–220.

David, A. and Ray, P., 1966. "Studies on the pollution of the river Daha (N. Bihar) by sugar and distillery wastes. *Evi. Hlth.*, 8: 6–35.

Greenfield, J.P. and Ireland, M.P., 1978. A survey of the macrofauna of a coal waste polluted Lanka Shireflurial system. *Environ. Poll.*, 16: 105–127.

Hovgaurd, P., 1963. A new system of sieves for benthic samples. *Sarcia*, 53: 15 18.

Janasson, P.M., 1955. The efficiency of sieving techniques for sampling fresh water bottom fauna. *Oikos*, 6: 183–207.

Kandibane, M., Raguraman, S. and Ganapathy, N., 2005. Diversity of aquatic Hemiptera in a rice ecosystem of Madurai, Tamil Nadu. *Indian J. Environ. and Eccoplan.*, 10(2): 489–492.

Kapoor, N.N., 1974. Some studies on the respiration of stone fly *Nymphs paragnelina* media (Walker). *Hydrobiologia*, 44: 37–47.

Kumar, A., 1980. Studies on the life history of dragon flies, *Ceriagrion caromandelianum* (Fabricus) (Oenagriidae odonata). *Rec. Zool. Surv. Ind.*, 76: 248–258.

Kumar, S., 1993. Insect communities of the high altitude lake Dashauhar, Himachal Pradesh, India. *J. Ecobiol.*, 5(4): 251–254.

Michael, R.G., 1968. Studies on the bottom fauna in a tropical fresh water fish pond. *Hydrobiologia*, 31: 203–230.

Mishra, N., 1985. The effect of sewage with special reference to aquatic insects in the river Kshipra. *Ph.D. Thesis*, Vikram University, Ujjain (MP).

Nebekar, A.V., 1971. Effect of high winter water temperature on adult emergence of aquatic insects. *Water Res.*, Pergamon Press, 5: 777–783.

Oomachan, L., 1988. Macrobenthos of lower lake, Bhopal. In: *Proc. Nat. Symp. on Past, Present and Future of Bhopal Lakes*, p. 27–31.

Reddy, R.S. and Pandian, T.J., 1972. Heavy mortality of *Gambusia affinis* reared on diet restricted to mosquito larvae. *Mosquito News*, 32(1): 108–110.

Roback, S.S., 1962. Environmental requirements of Trichoptera. In: *Biological Problems in Water Pollution, IIIrd Seminar*. Hlth. Serv. Publ. 999, W.P., 25: 118–126.

Saxena, M.N., Sharma, S. and Kaushik, S., 1989. Studies on aquatic insect communities and their distribution. *J. Hydrobiol.*, 5(1): 39–42.

Stoneburner, D.L. and Smock, L.A., 1979. Seasonal fluctuations of macroinvertebrate drift in a South Carolina piedmont Stream. *Hydrobiologia*, 63(1): 49–56.

Sridharan, S., Balasubramani, V., Jeyarani, S. and Sadakathulla, S., 2000. Aquatic hemiperan predators in rice ecosystem. *Insect Environment*, 5: 189–190.

Tonapi, G.T. and Rao, H.N.M., 1970. A biometrical analysis of growth in larvae of *Dineustes indicus*. Aube (Cryrinidae coleoptera). *Ind. J. Ent.*, 32(1): 39–50.

Chapter 13

Seasonal Variations in Insects Population of Lower Lake of Bhopal in Relation to Macrophytes

Madhulika Singh[1] *and K. Borana*[2]

[1]*Sadhu Vaswani College, Bairagarh, Bhopal*

[2]*Department of Applied Aquaculture, Barkatullah University, Bhopal*

ABSTRACT

In the present investigation, seasonal fluctuations have been noticed in the abundance of insects in relation to aquatic macrophytes. The Maximum density has been recorded in summer. Monthly density of total insects ranged from 111 to 488 nos/sqm in Lower Lake. The population density of Coleoptera, a dominant order in the lake, ranged from 16 to 237 nos/sqm, Diptera of 2 to 138 nos/sqm and Hemiptera 7 to 98 nos/sqm. The varying degree of association was observed between insects and macrophytes. The abundance of macrophytes exhibits a positive correlation with insect population.

Keywords: *Macrophytes, Insect population.*

Introduction

The insects attached to macrophytes have been of great importance as they affect the light conditions for plants and together with zoo benthos, contribute significantly to the nutrient cycling of transformation of organic matter. Different species of macrophytes support different macro invertebrates fauna due to their specific requirement of food, feeding habits and vegetation density. The plant surface area is also a criteria for macro invertebrate attachment as shown by several workers, submerged plants with fully dissected leaves offer better shelter than do the emergent macrophytes and submerged macrophytes with simple leaves.

The preference of insect to colonies certain plant species was shown by Velde (I960), Roweck (1988) and Chamber and Prepase (1990). Recycling of nutrients by macrophytes has been emphasised by many workers in their studies McRoy *et al.* (1972), Howard William and Junk (1976), Mickle and Wetzel (1978), Sridharan *et al.* (2000) and Kandibane *et al.* (2005).

Materials and Methods

The study of the insects and macrophytes were carried out in Lower lake of Bhopal from June 2004 to May2005. The mud samples were collected with Ekican Dredge simpler at monthly intervals. After that mud was sieved out through 0.5 mm mesh size sieved according to Jonasson (1955) and Hovgaurd (1963) methods. Samples of macrophytes were collected and identified with the help of guidelines given by Spence (1964) and Needham and Needham (1962). Correlation between insects and macrophytes were established.

Results

The insects were represented by three orders *viz.*, Coleoptera, Diptera and Hemiptera. The total insect density ranged from 111 org/sqm in September 2004 to 488 org/sqm in April 2005. Coleoptera was the most dominant and abundant order and it ranged between 16 org/sqm in December 2004 to 237 org/sqm in May 2005. Diptera was found dominant and abundant next to Coleoptera and it ranged between 2 org/sqm in July 2004 to 138 org/sqm in April 2005. The last dominant order was Hemiptera and it ranged between 7 org/sqm in June 2004 to 98 org/sqm in November 2004 (Table 13.1).

In the present study total twelve species of macrophytes were recorded. This include 3 species of emergent 5 of submerged, 1 of rooted forms with floating leaves and 3 of free floating forms, *Ipomea.*,

Jussiaea and *Polygonum* were the representative species of emergent forms, *Ceratophyllum, Hydrilla, Najas, Potamogeton* and *Vallisneria* were important species of submerged floating forms, *Nelumbo nucifera* was represented by rooted forms with floating leaves and *Azolla, Eichhornia* and *Leamna* were important species of free floating forms.

Table 13.1: Monthly Population Density (org/sqm) of Insects in Relation to Macrophytes from June 2004 to May 2005

Months	*Total Insects*	*Coleoptera*	*Diptera*	*Hemiptera*
Jun. 2004	324	218	12	07
Jul.	267	179	02	21
Aug.	227	143	34	57
Sep.	111	109	30	61
Oct.	201	89	39	84
Nov.	290	45	52	98
Dec.	272	45	65	81
Jan. 2005	310	65	89	52
Feb.	350	101	102	50
Mar.	450	155	79	47
Apr.	488	219	138	51
May.	417	237	84	40

Discussion

In the present studies insects groups were more prominent in the littoral zone where abundance of vegetation is available. David and Ray (1966) supported this pattern of abundance. Ray *et al.* (1966) also observed that in comparison to river Ganga and Jamuna possess a rich aquatic vegetation and fauna associated with the vegetation

In the present study the insects were found associated with 12 macrophytic species. The maximum number of macrophytes was observed during post-monsoon and winter season. Similar observations were also made by Penfound (1956). Insects in terms of number were seen maximum on *Eichornia, Hydrilla, Jussiae* and *Potamogeton.*

In the present study total insect density showed direct and significant correlation with macrophytes.

References

Chambers, P.A. and Prepas, E.E., 1990. Competition and Coexistence in submerged plant communities. The effect of species interactions versus abiotic factors. *Freshwater Biology*, 23: 541–550.

David, A. and Ray, P., 1966. Studies on the pollution of the river Daha (N. Bihar) by sugar and distillery wastes. *Env. Hith.*, 8: 6–35.

Hovgaurd, P., 1963. A new system of sieves for benthic samples. *Sarsia*, 53: 15–18.

Howard-Williams, C. and Junk, W.J., 1976. The decomposition of aquatic macrophytes in the floating meadows of Central Amazonian. Biogeograph., 7: 115–123.

Jonasson, P.M., 1955. The efficiency of sieving techniques for sampling fresh water bottom fauna. Oikos, 6: 183–207.

Kandibane, M., Raguraman, S. and Ganapathy, N., 2005. Diversity of aquatic Hemiptera in a rice ecosystem of Madurai, Tamil Nadu. *Indian J. Environ. and Eccoplan.*, 10(2): 489–492.

Me Roy, C.P., Barsdata, R, J. and Nebert, M., 1972. Phosphorus cycling in an eelgrosa (Zostera marina L.) ecosystem. *Limnol. Oceanogr.*, 17: 58–67.

Mickle, A.H. and Wetzel, R.G., 1978. Effectiveness of submerged angiosperm epiphyte complex on exchange of nutrients and organic carbon in littoral system. *Aquat. Bot.*, 4: 314–329.

Needham, J.G. and Needham, P.R., 1962. *A Guide of the Study of Freshwater Biology*. Holdeom Day Inc. San Francisco, pp. 108.

Penfound, W.T., 1956. Primary production of vascular aquatic plants. *Limnol. and Oceanogr.*, 1(2): 92–101.

Ray, P., David, A., Singh, S.B. and Sehgal, K.L., 1966. A study of some aspects of ecology of the river Ganga and Jamuna at Allahabad (U.P.), in 1958, 1959. *Proc. Nat. Acad. Sci. Ind*, B.36(3): 235–272.

Roweck, H., 1988. Okologische untersuchungen an Teichrosen. *Arch. Hydrobiol. Suppi.* 81: 103–358.

Spence, D.H.N., 1964. The macrophytic vegetation of Lochs, swanps and associated ferns. In: *The Vegetation of Scotland,* (Ed.) J.H. Burnett. Oliver and Boyd, Edinburgh, p. 306–425.

Sridharan, S., Balasubramani, V., Jeyarani, S. and Sadakathulla, S., 2000. Aquatic hemipteran predators in rice ecosystem. *Insect Environment,* 5: 189–190.

Velde, G.V., 1980. Studies in nymphacid-dominated systems with special emphasis on those dominated by *Nymphoides pellata* (Gmel.) O, Kuntze (Menyanthaceae). *Thesis,* Nijmegen, pp. 166.

Chapter 14

The Effect of Pollutants on the Genotype of Five Species of Family Coenagrionidae (Zygoptera : Odonata)

G.K. *Walia**

Department of Zoology, Punjabi University, Patiala – 147 002

ABSTRACT

Chromosomal analysis have been carried out on five species belonging to family Coenagrionidae *viz., Agriocnemis obscura* (♀), *Ceriagrion coromandelianum* (♂), *Coenagrion dyeri* (♀), *Pseudoagrion decorum* (♂) and *Pseudoagrion rubriceps* (♂ and ♀) collected from stagnant water bodies. Majority of plates possess the typical diploid number 27 in males and 28 in females with XO-XX sex determining mechanism. Numerous plates also show fragmentation of the autosomes and diploid number varies from 27-58. This type of chromosome variation is due to the effect of pollutants on the genotype of the species.

Keywords: *Autosomal fragmentations, Stagnant water bodies, Holokinetic chromosomes.*

---

* *Corresponding Author:* E-mail: gurinderkaur_walia@yahoo.co.in;
Phone: 01762-223720 (Res.), 09876020637 (Mob.)

Introduction

Dragonflies spend their entire developmental period in the freshwater environment. So the role of water is very important in the life cycle of odonate species. Now a days the entry of pollutants in aquatic system through industrial wastes containing lots of toxins and heavy metals affect the quality of water. This result in serious threat to the aquatic organisms including odonate larvae. Dragonflies emerged from polluted water bodies show the genotypic changes, which result in fragmentation of chromosomes.

Although in Odonata, chromosomal fragmentations are considered as normal mode of increase in chromosome number and responsible for numerical variations in all the karyotypes with more elements at haploid stage. So far, autosomal fragmentations have been reported by Cruden (1968), Kiauta (1969b, 1971a),Tyagi (1982) and Walia and Sandhu (1999c) in both suborders, Anisoptera and Zygoptera. During the present studies fragmentation of chromosomes have been observed in five coenagrionid species *viz., Agriocnemis obscura* (♀), *Ceriagrion coromandelianum* (♂), *Coenagrion dyeri* (♀), *Pseudoagrion decorum* (♂) and *Pseudoagrion rubriceps* (♂ and ♀) collected from vicinity of stagnant water bodies. The diploid number in such cells varies from 27-58. The specimens collected from fresh water bodies reveal diploid number 27m in males and 28m in females. These numbers are the typical coenagrionid numbers. The chromosomal data of fresh water specimens has been compared with the stagnant water specimens. This type of aberrant fragmentation is due to the effect of pollutants on the genotype of all the five species.

Materials and Methods

Odonate species were collected from North and North-East Indian states during the pre-monsoon and post-monsoon seasons. (Table 14.1).

The adult specimens of both the sexes were dissected in 0.7 per cent saline and chromosome preparations were made by air-drying technique as described by Sandhu and Walia (1994b). Structure and behaviour of chromosomes were studied during mitotic and meiotic cycle. Relative length of chromosomes was measured and compared within the species and other species of the same genus or family.

Table 14.1: Collection Data

Sl. No.	Name of Species	Locality	Place of Collection	Number of Specimens	Specimens Showing Fragmentation
1.	*Agriocnemis obscura*	Stagnant standing water bodies	Barapani (Meghalaya)	(♀)=3	3
2.	*Ceriagrion coromandelianum*	Stagnant standing and running water bodies	Jammu and Kashmir, Himachal Pradesh, Punjab, Uttaranchal, Assam and Meghalaya	(♂)=40	12
3.	*Coenagrion dyeri*	Stagnant standing water bodies	Nakuchia Tal (Nainital) Uttaranchal	(♀)=3	3
4.	*Pseudoagrion decorum*	Stagnant standing and running water bodies	Punjab, Bhim Tal, Nakuchia Tal (Nainital) Uttaranchal	(♂)=9	5
5.	*Pseudoagrion rubriceps*	Stagnant standing and running water bodies	Jmmu and Kashmir, Punjab, Assam and Meghalaya	(♂)=32 (♀)=5	9 2

Results

Majority of spermatogonial metaphase plates (70 per cent) and oogonial metaphase plates (60 per cent) showed the typical coenagrionid number that is 27 in males of *Ceriagrion coromandelianum, Pseudoagrion decorum* and *Pseudoagrion rubriceps* and 28 in females of *Agriocnemis obscura, Coenagrion dyeri* and *Pseudoagrion rubriceps* (Table 14.2). The diploid number 27 is differentiated into 26 autosomes and single X chromosome and 28 into 26 autosomes and two X chromosomes. These results demonstrate unequivocally the XO-XX sex determining mechanism. Autosomes also include pair of m chromosomes. In *Agriocnemis obscura* and *Coenagrion dyeri* m and X chromosomes could not be differentiated because males of these species were not found during

Table 14.2: Numerical Variations in Chromosome Number

Sl.No.	Name of Species	Sex	Total Number of Cells Studied	Chromosome Numbers													
				27	28	34	36	37	41	42	43	44	45	46	52	56	58
1.	*Agriocnemis obscura*	(♀)	61	0	42	0	3	0	0	0	0	5	0	0	7	4	0
2.	*Ceriagrion coromandelianum*	(♂)	8	40	0	0	0	0	6	0	0	0	2	0	0	0	0
3.	*Coenagrion dyeri*	(♀)	51	0	32	0	9	0	0	0	0	0	0	0	0	7	3
4.	***Pseudoagrion decorum***	(♂)	50	30	0	3	0	0	0	3	0	0	0	0	2	5	1
5.	*Pseudoagrion rubriceps*	(♀)	48	0	29	0	0	0	0	8	0	0	0	11	0	0	0
		(♂)	56	40	0	0	0	10	0	0	4	0	2	0	0	0	0

collection and therefore not possible to compare the data related to m and X chromosomes as well as sex determining mechanism.

The remaining spermatogonial metaphase plates (30 per cent) and oogonial metaphase plates (40 per cent) showed the fragmentation of autosomes. The number varies from 30 to 58 in fragmented spermatogonial metaphases and 36 to 58 in fragmented oogonial metaphases. The X and m elements could not be recognized in such fragmented metaphase plates.

Males of *Pseudoagrion decorum* and *Pseudoagrion rubriceps* reveal the fragmented chromosomes at metaphase-I stage with 19 elements, might be because of some of the fragmented spermatogonial metaphase stages entered into the meiosis. However, beyond metaphase-I no subsequent stages were found.

Discussion

Kiauta (1969) presented an elaborate account of the autosomal fragmentations and fusions in Odonata with special reference to their evolutionary implications. According to him, chromosome fragmentations lead to increase in the chromosome number during the evolution of this order. Cruden (1968) and Kiauta (1969b) in *Enallagma cyathigerum* (n=14/15). Kiauta (1969b, 1971a) in *Diplacodes bipunctata* (n=12/13), *Libellula depressa,* (n=12/13). *Orthetrum coerulescens* (n=12/13) and *Calopteryx virgo meridionalis* and Tyagi (1982) in *Ceriagrion cerinorubellum* (n=14/15) and *Anisopleura testoides* (n=13/14) noticed slight increase in the haploid number of these species and supported Kiauta's view. During the present study, the typical coenagrionid number that is, in males 2n= 27m and in female 2n=28m was found in the specimens collected from fresh water bodies, while autosomal fragmentations have been observed in those specimens collected from the stagnant water bodies. As odonate spend their entire developmental period in the freshwater environment. So the role of water is very important in their life cycle, but due to the entry of pollutants in aquatic system through industrial wastes containing lots of toxins and heavy metals affect the quality of water. This result in serious threat to the aquatic organisms including odonate larvae. Adult dragonflies emerged from the larvae of stagnant water bodies show the genotypic changes, might be due to the effect pollutants present in the water. The chromosomal fragmentations observed in the five species of family Coenagrionidae is, therefore resulted due to effect of pollutants, having no bearing on evolutionary trends of this order as these aberrations were observed only in 40 per cent of the cells (Table 14.2)

It was interesting to note that some of the fragmented gonial cells entered into meiosis, which showed paired fragmented chromosomes at metaphase-I plate and behaved normally as the normal metaphase chromosomes. It seems to be merely the holokinetic nature of the odonate chromosomes.

Acknowledgements

I am thankful to DST New Delhi for financial assistance and A.R. Lahiri for the identification of specimens.

References

Cruden, R.W., 1968. Chromosome number of some North American dragonflies (Odonata). *Can. J. Genet.Cytol.*, 10: 200–214.

Kiauta, B., 1969b. Autosomal fragmentations and fusions in Odonata and their evolutionary implications, *Genetica* 40: 158-180.

Kiauta, B., 1971a. An unusual case of precocious segregation and chromosome fragmentation in the primary spermatocytes of the damselfly, *Calopteryx virgo meridionalis* Selys, 1873, as evidence for a possible hybrid character of some populations of the *Calopteryx virgo* complex (Odonata, Zygoptera, Caloptrygidae) *Genen Phaenen*, 14(2): 32–40.

Sandhu, R., and Walia G.K., 1994b. Chromosome studies of three species of libellulids (Anisoptera: Odonata). *La Kromosome* II 75–76: 2599–2604.

Walia, G.K. and Sandhu, R., 1999c. Autosomal fragmentations in five species of family Coenagrionidae (Zygoptera: Odonata). *Fraseria*, 5: 15–20.

Tyagi, B.K., 1982. Cytotaxonomy of the Indian dragonflies. *Indian Rev. Life. Sci.*, 2: 149–161.

Chapter 15

Main Cytogenetic Characters in the Order Odonata

G.K. *Walia*

Department of Zoology, Punjabi University, Patiala – 147 002

Odonata are the best cytologically studied group among insects (Kiauta, 1975 and Tyagi, 1982). These insects are unique in presenting cytological uniformity on the one hand and sharp cytological specificity on the other. According to Kiauta (1975), no other known insect group exhibits major karyotypic variation and deviates at the same time to such an extent from all other orders of the class Insecta. The karyotypic variation may perhaps be due to the fact that there are extraordinarily long phyletic lines. However, the deviation of distinct cytological characters of dragonflies are certainly due to the ancient origin and isolated phylogenetic position of the order within the living insects.The cytological study of order Odonata can be characterized under five major cytogenetic characters, which include:

1. Centromere
2. Recombination index
3. Chromosome number
4. m chromosomes
5. Sex chromosomes and sex determining mechanism

Centromere

The subject of the centromeric condition in order Odonata has always been a matter of controversy. Oksala (1943b, 1944b, 1945, 1952) was the first to describe the localized centromere in this group. He found metacentric chromosomes with arms of exactly the same length. His conclusions were based on two evidences. Firstly, in polar view of spermatogonial metaphases, the chromosomes were often bent in the middle region and therefore the centromere was possibly located at this constriction. Secondly, in lateral view of some primary spermatocyte metaphases, the two spindle fibres appeared to connect each chromosome with each pole.

Subsequently, Lima de Faria (1949), after studying Oksala's papers, suggested no localized centromere in odonates. His conclusions were exclusively based on Oksala's observations as he himself did not study any material. According to him, in some species the bivalents at primary spermatocyte metaphase had their longitudinal axes placed at right angles to the spindle and from some bivalents more than one chromosomal fibre could be running towards the same pole. This view was shared by Hughes-Schrader (1948), while studying the cytology of coccids and also by some other workers, but none of them produced any new evidence.

Piza de Toledo (1950, 1953), while elaborating his views on the status of centromere in dragonflies and in Hemiptera, reported the chromosomes to be dicentric. He based his assumption on the basis of the chromosomes appearing curved at both ends towards the poles. However, he too did not study any original material.

Cumming (1964a) proposed diffuse nature of the centromere in Odonata for the first time. His observations were based on 109 species of 11 families of Odonata. His inferences were drawn on the discovery of species with strickingly low chromosome numbers and also on the observations of large-sized chromosomes in low (n) species during spermatogonial and primary spermatocyte anaphases proceeded to the poles at a parallel manner. Inspite of the large size of chromosomes in low (n) species, no structure appeared which could be interperated as a median centromere. On the other hand, the spindle fibres were attached to the whole length of the chromosome. He further concluded that all the chromosomes of low (n) species were at least twice the length of those of closely related high n species. The phenomenon, according to him, was similar to the situation met within other insect orders possessing diffuse centromere. He also studied phase contrast pictures of Feulgen squash preparations and observed two spindle fibres proceeding to each pole from each univalent during the spermatogonial anaphase and also from each dyad at secondary spermatocyte anaphase. These fibres were not close enough to be interperated as a compound spindle fibre from a localized centromere. He suggested that these represented the borders of regions of kinetic activity of a chromosomal element or the optically observable edges of a broad band of spindle fibres attached along the entire length of the chromosome.

Cumming's views regarding the diffuse nature of centromere were subsequently strengthened by Kiauta (1967e, 1968d, 1969a,b, 1975) on cytotaxonomic and phylogenetic grounds. He based his conclusions on comparison of morphological and ethological features and of phylogenetic trends in the odonate karyotype, with those of other organisms possessing holocentric chromosomes. He gave some experimental evidences, in which he discussed the general appearance of chromosomes and the orientation of elements in polar view of metaphase plate. From his observations he inferred that chromosome fusion was evidently responsible for the reduction of chromosome numbers and did not affect the amount of DNA in the secondary set, which remained equal to the DNA of the primary complement. On the other hand, the increase in chromosome number paralleled to specialization and/or advanced phylogenetic position of the forms involved. The chromosomal fragments originated by spontaneous fragmentation of one or more elements in some or all cells in atleast some natural populations of some species survived in mitosis, *viz.* in *Enallagma cyathigerum*. *Meicstogaster* species, *Libellula depressa, Orthetrum coerulescens, Diplacodes bipunctata, Diplacodes haematoides* and *Neurothemis tullia*. He argued, these fragmentations would have disappeared in the course of mitotic division if the chromosomes were monocentric. Another support of his views was that the broken parts of the irradiated mitotic elements behaved normally or exhibited normal kinetic behaviour.

Kiauta (1970a) emphasized the fact that most of the phylogenetically old orders of insects possessed holokinetic chromosomes, same way Odonata also fitted in the general scheme. Although Ephemeroptera the nearest allies of Odonata, did not possess the holokinetic chromosomes but had monominetic chromosomes. Holokinetic nature of chromosomes in Odonata has generally been accepted universally, but few workers had also recently offered different and contrasting opinions.

Gama *et al.* (1981) advocated polycentric nature of odonate chromosomes on basis of the ultrastructural study of *Enallagma cheilferum*. Handa *et al.* (1984) studied the chromosomes of ten species of dragonflies and observed the presence of chiasmata during meiosis-I giving square to rectangular and other shapes to the bivalents,.They suggested that the renewed activity of kinetochore became localized during metaphase-I. Thomas and Prasad (1984), however, believed that Odonata definitely possessed a localized centromere.

The differential staining technique of C-banding has made it possible to stain constitutive heterochromatic regions at condensed stages of the division cycle, when such regions are normally indistinguishable from euchromatin. Thomas and Prasad (1986) and Suzuki and Saitoh (1988) Perepelov *et al.* (2001) and Walia (2006) made preliminary investigations regarding the distribution of constitutive heterochromatin in Odonata. They found that there was a peculiar distribution pattern in sex chromosomes, while autosomes did not show any differential staining. C-bands were present only at the terminal ends of autosomal bivalents. From these observations it is depicted that the odonates possess holocentric chromosomes.

Recombination Index

Kiauta (1975) in his exhaustive communication on the taxonomic organization of the order and representativeness of the cytologically investigated data had stated quite different evolutionary patterns in species with low recombination index (RI) and high recombination index. Low RI promoted fitness that is, the survival value and the reproductive capability of a genotype as compared to the average of the population or the other genotypes in it. On the other hand, high RI related to high flexibility, that is, the ability of the variable genotype and its adaptation to the changing conditions. To maintain a balance between genetic variability, biological efficiency and stability, the organisms evolved their recombination indices, which optimally suited their evolutionary requirements.

He further added occurrence of only single chiasma per bivalent in male, whereas two chiasmata per bivalent in female dragonflies and found recombination index 2n-lin (XO) males, while 3n in (XX) females. Evidently, the chiasma frequency was almost fixed in the order, so any change in chromosome number was the only way of changing the species recombination index. The presence of small numerical variation of the complement within odonate families suggested firstly, recombination indices had been stabilized at selectively adaptive levels of the families concerned. Secondly, relatively few species of a family had a need to deviate from the general pattern adapted by the family. The later applied both to the primary and secondary complements.

In odonates, chromosomes did not show any visible structural differentiations; only those rearrangements of the karyotypes could be studied, which produced numerical increase or decrease of the chromosome complement (numerical variations). This resulted in the alteration in the gene sequence and most probably to the subsequent position effects. Kiauta (1975) concluded that chromosomal fusions and fragmentations played a very important role in the evolution and phylogeny of the order. These views were subsequently supported by Tyagi (1982, 1986).

In families Platycnemididae and Calcinemididae, Kiauta and Kiauta (1981) reported that recombination index in males was stabilized at 2n–1 = 24. However, they observed a low(n) number in *Risiocnemis incisa* belonging to the former family, resulted in the reduction in recombination index that is, from 24 to 20. This favoured the survival value and reproductive capability of the species and made it fit to settle over a wide geographical range. However, the disadvantage lied in the restriction of its ability to adapt to substantial ecological diversity of habitats.

A single chiasma per bivalent observed in the males of Odonata. However, rare examples were met with where two chiasmata per bivalent were present. Ferreira *et al.* (1979) in *Orthemis ferruginea* and *Erythrodiplax umbrata* claimed the occurrence of two chiasmata per bivalent in atleast 8 bivalents. Mola and Agopian (1985) also observed two chiasmata per bivalent during early diakinesis in males of *Tauriphila risi* and *Perithemis mooma*. However, in the discussion they contradicted the statement and wrote that "assume single chiasma occurs as a rule".

Walia (1997) studied 60 male species of order Odonata and reported single chiasma per bivalent during the diplotene/

diakinesis stages, thus indicated that the species were stabilized in their habitat "at selectively adapted levels of the family" as suggested by Kiauta (1975). The numerical variation that result in any changes in number due to fragmentations or fusions were few. Autosomal fragmentations were observed only in five species, while majority of the cells possessed normal chromosome number (Walia and Sandhu, 1999c).

Chromosome Number

White (1948) gave 2n = 25 as the type number in dragonflies, as it was the most commonly found number in suborders Zygoptera and Anisoptera. Dasgupta (1957) was also of the opinion that the most frequent number of chromosomes in a peculiar systematic group ought to be termed as the type number. He, however, considered the type number to be the most primitive chromosome number of group concerned and viewed that other numbers found in a group were normally derived from this type number.

Kiauta (1975), while reviewing the taxonomic organization of order Odonata reported that male haploid chromosomes numbers in dragonflies ranged from 3-15. However, inspite of this rather broad range, there was in general, little numerical variation in dragonfly complements. As haploid numbers 12, 13 and 14 were represented in more than 95 per cent of Odonata examined, hence n = 13 could be considered as the type number of the order as was earlier suggested by White (1948).

At the suborder level, the pattern was found to be almost uniform in Anisoptera. The haploid number, n = 13, was represented in 70 per cent of species studied. The same type number could also be assumed for Anisozygoptera. This suborder is represented by only one cytologically studied species (Dasgupta 1957, Kiauta 1975, Tyagi 1982). In suborder Zygoptera, however, more variations were found, thus complicating the situation. The haploid number n = 13, was found in less than 50 per cent of cases, while in majority of the species, 14 elements occurred in the haploid set. From this evidence, n = 14, should be regarded as the type number of the suborder. However, this number was peculiar only to two families, Protoneuridae and Coenagrionidae and the later family covered more than half of the total number of representatives of the suborder examined cytologically. Kiauta (1975), on this basis, suggested n = 13 as the type number for the suborder Zygoptera. Walia (1997) also reported that out of 60 odonate species 2n = 25 was found in 34 species, the only exception was the family Coenagrionidae with 2n = 27in16 species. She concluded 2n = 25 to be the type number of the order Odonata.

Dasgupta (1957) reported that chromosome number fluctuated not only at the generic level but also among the species of the same genus. He quoted the species of genus *Sympetrum* in which *eroticum* 2n = 21, *frequence* 2n = 23, while the remaining species possessed 2n = 25. He also observed similar variations among the genus *Gomphus* in which *Gomphus hakiensis* and *Gomphus suzukii* possessed 23 chromosomes, while *Gomphus unifasciatus* and *Gomphus melampus* showed the diploid numbers 21 and 19 respectively.

Two types of variations among chromosome numbers were observed in the odonate chromosomes. These were, variation of chromosome numbers within the same species and variation of chromosome numbers within the cells of the same specimen. Congeneric variation in chromosome numbers was reported by Dasgupta (1957) in *Aeshna coerulea* possessing 2n = 25, which was deviated from the typical number 27 found in other species.

Kiauta (1969b) described some genera deviated distinctly from the original chromosome type number of the families to which they belong *viz. Leptagrion* (n = 15) in Coenagrionidae, *Mecistogaster* (n = 13 and n = 15) in Pseudostigmatide, *Trigomphus* (n = 10 and 12) in Gomphidae, *Tetragoneuria* (n = 11 and 14) in Cordulidae and *Orthemis* (n = 12) in Libellulidae. According to Kiauta (1975), some of these complements were secondary in origin.

In most of the species, the chromosome number remained constant but in some instances more than one chromosome number was reported. Such variation occurred either in different cell of the same specimen or in different geographic populations of the same species. Kiauta (1975) reported such a situation in *Trithemis aurora* collected from Nepal in which he found n = 10 and 11 in the same individual at primary spermatocyte metaphase. In *Rhinocypha quadrimaculata* he found n = 12 occurring in all cells and in all individuals collected from Nepal but observed n = 13 in material collected from Darjeeling.

Another phenomenon of interest among odonate chromosomes was the precocious segregation of chromosomes leading to aberrant variation in chromosome number. Kiauta (1971a) reported the variation in chromosome number in *Calopteryx virgo* due to precocious segregation of chromosomes. Walia (1997) also studied same variation in *Ceriagrion coromandelianum* in which the smallest autosomal bivalent (m chromosomes) appeared to be precociously segregated at the diakinetic plate.

Kiauta (1969b) discussed that changes in chromosome number in dragonflies could be due to autosomal fragmentation or autosomal fusion of the chromosomes. Chromosomal fragmentation was considered to be the normal mode of increase of chromosome number in the course of evolution of the Odonata, and the only way by which recombination index was increased. The selective adaptive level of the recombination index was uniform in family, only few species deviating from the family type number. Fragmentation was observed only in some specialized and advanced dragonflies, which were capable of retaining their flexibility by increasing the recombination index, otherwise they would be unable to adapt themselves to the changing conditions as being specialized to survive in the altered conditions. Thus, it confirmed the principle of parallelism between the increase of specialization and that of chromosome number. The fragmented element behaved just like the normal chromosome and generally had the form of the 'm' chromosome, irrespective of the presence or absence of the 'm' chromosome in original complement.

He observed the fragmentation of autosomes in the karyotype of *Enallagma cyathigerum* (n = 14-15), *Mecistogaster* species (n = 15). *Hetaerina rosea* (n = 14), *Libellula depressa* (n = 12-13), *Orthetrum coerulescens* (n = 12-13), *Diplacodes bipunctata* (n = 13-15) and *Diplacodes haematoides* (n = 12-13). Tyagi (1978b) also reported chromosomal fragmentation in *Ceriagrion cerinorubellum* (n = 14-15) and *Anisopleura testoides* (n = 13-14). Walia and Sandhu (1999c) observed an aberrant increase in chromosome number in five zygoptern species belonging to family Coenagrionidae *viz.*, *Agriocnemis obscura* (2n ♀ = 28 more than 50), *Ceriagrion coromandelianum* (2n ♂ = 27 = more than 50), *Coenagrion dyeri* (2n ♀ = 28 more than 50), *Pseudoagrion decorum* (2n ♂ = 27–more than 50) and *Pseudoagrion rubriceps* (2n ♂ & ♀ = 27–more than 45). Such a variation could only be due to the fragmentation of chromosomes. Moreover, normal behaviour of the fragmented autosomes through the mitotic and meiotic stages also substantiated the holokinetic nature of odonate chromosomes, which had been a subject of controversy (Oksala 1943b, 1944b, 1945, 1952, Lima de Faria 1949, Piza de Toledo 1950, 1953, Cummings 1964a and Kiauta 1975).

Kiauta (1967e) stated that chromosome fusions were responsible for the decrease of chromosome number in the secondary complements and were not related to the phylogenetic status of the taxa concerned. Kiauta (1969a) substantiated the above statement and observed essential difference between the fusion of an autosome with sex element and the fusion of two or more autosomes. The former fusions were characterized by the appearance of both fused and unfused complements in one individual, whereas the later fusions were specifically characteristic and occurred in all cells, all individuals and all populations of the species. It is a fact that the complement after fusion was secondary in origin because many dragonfly complements having low number than type number were primary rather than secondary in origin. He further elaborated the two features, which were characteristic for the complements reduced secondarily by an autosomal fusion, which were not found in primary complement, within a genus in which one or some species had low chromosome numbers than type number of the group. Karyotype could be assumed to be of secondary, rather than of primary origin. The secondary complement was recognized by one large autosomal bivalent and also more number of chiasmata per bivalent.

Autosomal fusions were earlier reported by Cumming (1964a) in *Aeshna diffinis diffinis* (2n ♂ = 21, n = 11), *Aeshna intricata* (2n ♂ = 19, n – 10) and *Perithemis lais* (2n ♂ = 17, n – 9). Such fusions probably also caused numerical reduction in *Lestes forcipatus* (n = 11, Cruden 1968). Kiauta (1969b) observed diploid number 21 in *Orthetrum brachiale, Sympetrum eroticum eroticum* and also in *Risiocnemis incisa* (Kiauta 1981). Tyagi (1978b) also reported fusions in both zygoptern and anisoptern dragonflies.

From the above data Tyagi (1986) drew two conclusions: (1) The low 'n' species had one large autosomal pair, (2) The chiasma frequency was higher in secondary low 'n' species. In this way the recombination index in primary and secondary complements remained the same or may become slightly increased in secondary complement. Wherever, the sex element was involved in fusion with one or more autosomes it evolved a neo-XY sex determining mechanism. Fusion favoured the survival value and reproductive capability of the species to settle down over the whole geographic range.

Evolution of Odonate Chromosome Numbers

Oksala (1943b), Seshachar and Bagga (1962) and Cumming (1964a) discussed the evolution of odonate chromosome numbers and were of the opinion that chromosome numbers lower than type number had come about by fusion of the elements of ancestral karyotype. A different theory, "the 'm' chromosome theory", had been proposed by Oguma (1930) and supported by Dasgupta (1957), according to which the 'm' chromosomes were infact autosomes undergoing a gradual diminution in volume until they eventually disappeared one after another. The intermediate stage during the process of diminution was indicated by the presence of chromosomes ('m' chromosomes), which were distinctly smaller in size when compared to the other autosomes. Dasgupta (1957) considered the complement n = 14 (13a + X), the highest chromosome number known at that time, as the ancestral number. Thus according to him, the families possessing this type number were the most primitive families such as Coenagrionidae and Aeshnidae.

The above theory, however, was criticised by Kiauta (1968c, 1975) and Tyagi (1986) who shared a different view. They pointed out that the families with high 'n' number were infact the more advanced ones and assumption of gradual diminution of 'm' chromosomes would lead to the loss of DNA material from the chromosomal complement, which was not feasible.

Kiauta (1967e) suggested that n = 9 was the ancestral number of the odonates and divided the dragonfly karyotypes into two groups. One was of the normal high 'n' complements (n = 9-15), while the other was of the low 'n' complements (n = 3-7). Among high 'n' complements, n = 9 was the lowest number and was possibly the most probable primitive ancestral chromosome number. He further opined that on paleontological grounds, it could be suggested that this number represented the true ancestral number of the order as it was found in the primitive families, Pseudolestidae and Petaluridea, which were more archaic and phylogenetically most primitive among living dragonflies. Another primitive and early side line was the family Gomphidae, in which chromosome numbers in many species deviated from the supposed ancestral number. Among more specialized forms, only one case of n = 9 was known in *Perithemis lais,* which belonged to family Libellulidae. On basis of these reasons he assumed that the complement n = 9 represented the most probable ancestral chromosome number of the order.

Numerical variation in odonates was discussed in detail by Kiauta (1967e, 1969b, 1975). According to him karyotypes developed through the occurrence of breaks or fragmentation leading to n = 10 to 15 and fusions leading to n = 3 to 7 in ancestral forms. All these, as mentioned earlier, had the basic chromosome number n = 9. Fusions of two or few elements give rise to secondary high n types (n = 14-13-12 etc.) In such cases fusions could be traced because fused chromosomes were longer than other elements of the set. The author further indicated that if the fusion involved only two chromosomes, one was the sex chromosome that resulting in neo-XX-XY sex determining mechanism, which replaced the usual XX-XO type. Fusion was not always found in all populations of the same species nor did it always occur in all cells of one individual.

Kiauta (1967e, 1969b, 1975) justified the advantage of this hypothesis of karyotypic evolution in dragonflies based on the integration of independent karyological, morphological and paleontological evidences and accounted for the origin of any chromosome number.

'm' Chromosomes

Lefevre and McGill (1908), while studying the cytology of *Anax junius,* were the first to use the name micro chromosomes or 'm' chromosomes for a pair of minute chromosomes occurring in this species. The nomenclature was rapidly adopted by other odonate cytologists, particularly in Japan, where most of the early cytological work on dragonflies was carried out. This term is now used regularly in the literature. Very little is known about the nature of 'm' chromosomes in Odonata and there has been some controversy whether or not these chromosomes should be called 'm' chromosomes like those present in family Coreidae. Makalovskaja (1940) was the first to object to this nomenclature on the basis of lack of behavioural differences between the 'm' chromosomes and other dragonfly autosomes. Subsequently, Srivastava and Das (1953) too shared her opinion.

Kiauta (1968b,c) acknowledged the presence of m chromosomes as they were present in 80 per cent of the odonate species and were distinctly smaller than the other autosomes. Their occurrence was not bound to certain genera or infrageneric group of species. Closely related species could often be distinguished by the presence or lack of the 'm' chromosomes. Moreover, their relative and absolute size varied from species to species, within single species and geographical population and also among various cells of the same stage in the same individual.

Oguma (1930), Dasgupta (1957) and Cumming (1964a) proposed the 'm' chromosome theory of the karyotypic evolution, considered dragonfly 'm' chromosome to be an autosome undergoing gradual dininution in volume until it eventually disappeared. On the other hand Kiauta (1968c) considered the dragonfly 'm' chromosomes to be fragments of normal autosomes. He argued that only few closely allied (sub) species were known with the presence or absence of 'm' chromosomes possessing the chromosome numbers $n = n_1a + m + X$ and $n = n_1a + X$ respectively. However, if gradual disappearance of the 'm' chromosomes played any role in speciation, then (sub) species showing the presence of 'm' chromosomes should be more commonly found. Earlier, Hogben (1921) had also reported diploid chromosome number 2n = 23 without 'm' chromosomes in *Libellula depress* L collected from England. Later, Kiauta (1968c) collected the same species from Austria and demonstrated the presence as well as absence of 'm' chromosomes among the cells of the same individual. Thus, it could be presumed that in this species karyotypic evolution of 'm' chromosome is in progress. Kiauta (1968b) presented the cytological data of several widely distant populations of *Calopteryx virgo* throughout the zoo geographical range, while going from West to East, he noticed a change in the size of 'm' chromosomes. Population collected from Leningrad (West) there were no 'm' chromosomes, population from Japan (East) had minute 'm' chromosomes, while in Austrain and Solvene populations the 'm' chromosome was of intermediate size.

Tyagi (1986) suggested that the absence or presence of the 'm' chromosomes in any species was helpful in determining the taxonomic status of the taxon concerned. He quoted *Sympetrum commixtum* among all the libellulid species, in which 'm' chromosomes were absent. However, Walia (1997) reported presence of 'm' chromosomes in the same species. This might be due to geographical variation in chromosome number because Tyagi's collection was from Dehradun Valley (Uttaranchal), while the Walia's collection was from Patnitop (Jammu and Kashmir) located North West to Dehradun at high altitude. Similarly, Chatterjee and Kiauta (1973) reported 2n = 25 in *Rhinocypha quadrimaculata* from Darjeeling, while Kiauta (1975) reported 2n= 23 'm' in the same damselfly from Nepal. Regarding the relative size of 'm' chromosomes, Tyagi (1986) stated that it might be different in different infraspecific forms of the same species as was found in *Ischnura senegalensis, Orthetrum pruinosum neglectum* and *Pantala flavescens.*

Walia (1997) also reported presence or absence of 'm' chromosomes in various species *viz., Ischnura aurora aurora,* and *Ischnura forcipata* had diploid chromosome number 2n = 25, without 'm' chromosomes, while Kiauta (1975) and Tyagi (1978b, 1982) reported 2n = 27 for the same species. Similarly, in *Orthetrum japonicum internum, Orthetrum iuzonicum, Tholymis tillarga* and *Trithemis festiva* possessed 2n = 23, without 'm' chromosomes, while Omura 1955; Kiauta 1975; Kiauta and Ochssee 1979, Boyes *et al.,* 1980; Kiauta and Kiauta 1982, Tyagi 1978b, and Handa *et al.,* 1984 reported 2n=25m on the same species. These 'm' chromosome variations were due to the geographical variations because the place of collection was different for the same species.

Sex Chromosomes and Sex Determining Mechanism

White (1957), Saez (1963) and Darlington (1965) proposed XY/XX sex determining mechanism was the most primitive in organisms and all other modes of sex determination were derived from this type. Kiauta (1968c) was of the opinion that in Odonata, XO-XX is the most primitive type of sex determining mechanism, while neo-XY mechanism had secondarily originated from this type. The neo-XY sex determination occurred in those cases where original X was involved in fusion with an autosome and this condition was often reversible.

McGill (1904) was the first to notice the sex element in *Anax junius* and named as the "accessory chromosome". She was not able to recognize its proper sexual nature, nevertheless she described and figured correctly its post-reductional behaviour during meiosis. Lefevre and McGill (1908) were again the first to suggest a possible sexual character of the "accessory chromosome" in *Anax junius*, while Smith (1916), finally recognized the sex element in *Sympetrum semicinetum* (Say).

The neo-XY-XX mode of sex determination was discovered in dragonflies by Makalovskaja (1940). She properly described and figured the situation in *Aeshna grandis* and *Aeshna juncea* but she was not able to understand the mode of sex determination. Oksala (1943b) while studying the neo-XY sex determination in three species of *Aeshna*, stated that such systems involved an obligatory fusion of the original sex chromosome with an autosome. Later, Ray Chaudhury and Dasgupta (1949) in *Neurothemis tullia tullia* reported neo-XY mode of sex determination originating in a unique way. There were 28 chromosomes at spermatogonial metaphase, whereas 13 bivalents (including the 'm' bivalents) and unpaired neo-X and neo-Y occurred in primary spermatocyte. The later did not pair at diplotene and diakinesis. The first maturation division was equational for the sex element, while they lay in a forward position on either side of the equational plane in the second anaphase. *Neurothemis tullia tullia* was the only libellulid species, which had higher chromosome number than type number which was 2n = 25. Kiauta (1968d) proposed that numerical increase could be due to the fragmentation of two autosomal pairs and subsequent fusion of the original X with a fragment of an autosome. The fragmentation resulted in a diploid complement 2n = 29 in which XO sex determination was still retained. The subsequent fusion of original X with an autosomal fragment reduced the chromosome number by one and gave rise to neo-XY sex determining system. *Neurothemis tullia tullia* was thus the only dragonfly in which neo-XY condition did not originate in a fusion of the unpaired X, nor with an entire autosome but only with a fragment of an autosome. Subsequently, Handa *et al.* (1984) reported XO sex determining mechanism in this dragonfly, which, however, was collected from a different locality.

In *Pesudothemis zonata*, Omura (1955) reported an obligatory fusion of sex chromosomes in the same manner as described earlier by Oksala (1943b), while in *Crocothemis servilia* both neo-XY and XO type sex determining mechanism were found. However, in an earlier report by Asana and Makino (1935) only XO sex determining mechanism was noted.

Seshachar and Bagga (1962) observed diploid chromosome number 14 both in males and females of *Hemianax ephippiger*. The sex determining mechanism was of XX-XY type. The X was largest among the complement and was markedly bent. The Y chromosome was medium sized and rod shaped. The decrease in chromosome number was suggested to be possibly due to centric fusion of autosomes. If this was so, then *Hemianax ephippiger* must have acquired at least 13 centric fusions, so that its diploid number reduced from 27 to 14. One fusion could be between X chromosome and an autosome. Later Cruden (1968), while conducting the cytological studies on some North American dragonflies, supported the neo-XY system in *Ophiogomphus bison, Epitheca cynosura* and *Leucorrhinia frigida* but also observed that this system was not obligatory for all meiotic cells.

Kiauta (1969a, 1975) and Tyagi (1986) explained that in dragonflies the increase in advancement and specialization was accompanied by a numerical increase of the chromosome number which came about exclusively by successive autosomal fragmentations, while the orignal XO condition remained unaffected. Not a single case of fragmentation of original sex element had ever been reported in dragonflies. In the primary complements therefore, the XO-XX was the only mode of sex determination. On the other hand, in Heteroptera, which was also characterized by the diffuse condition of the centromere, the fragmentation of original sex element had been reported in numerous species belonging to atleast ten different families.

They further elaborated the process of stepwise numerical reduction. As far as the sex element was concerned, it was a cycle of three successive stages. In the first one, one autosome of the primary complement fused with the original X, which resulted in neo-XY karyotype, leading to the numerical reduction of the primary complement by one. In the second stage, fusion of neo-Y with an autosome occurred, which reduced the chromosome number by another element and gave rise to a neo-X and neo-neo-Y sex determining mechanism. The final stage was achieved by translocation of the autosomal part of neo-X to an autosome. This resulted in the reappearance of XO system, which was not accompanied by a numeric reduction. All three stages were demonstrated in the family Gomphidae, and relatively common occurrence of neo-XY complement in most advanced families suggested the universality of the process. They also reported that the reduction was not stable or obligatory and could be reversed at any time to reach the state of original XO type of sex determining mechanism. So both types of sex determination might be met within the same species among specimen collected from different location as was mentioned earlier.

From the above discussion it could be inferred that XO was the primary sex determining mechanism in Odonata. The secondary complements originated by fusion of the elements of primary chromosomal set and did not have any relationship with the phylogenetic position of the species. Thus both XO-XX and neo-XY sex determining mechanisms occurred in dragonflies.

A variation in the size of the sex chromosomes is another prominent feature of Odonata. Kiauta (1975) reported that among the species having XO sex determing mechanism, the X was usually the smallest element in the whole complement, though in few cases it was of medium size and it could also be the largest. Dasgupta (1957) observed size of the X chromosome remained constant during cell division and reported that the m : X ratios, in this order, were unchanged for a given species in all the meiotic stages.

White (1973) reported that in many groups of insects with holokinetic chromosomes the meiotic process was either pre-reductional or post-reductional. In Odonata, however, the type of meiosis was a subject of controversy, and two schools of thought exist. Seshachar and Bagga (1962), Handa *et al.* (1984) and Mittal and Gandhi (1984) described pre-reductional meiosis, while Oksala (1943b), Cumming (1964a), Cruden (1968) and Kiauta (1969a, 1975) found post-reductional division of the X element. Mola (1995) explained that these two lines of thought could be a consequence of the difficulty to determine the mode of orientation (axial or equational) of bivalents at metaphase I. The high degree of condensation resulted in close aggregation at equational plate, and in many species, the high number and small size of bivalents makes it difficult to determine the exact placement of the elements.

Pre-reductional division of the X chromosome in *Libellula luctosa* (anisoptera), and in *Enallagma cyathigerum* (zygoptera), was reported by Smith (1916) and Makalovskaja (1940) respectively. Oksala (1943b) made analysis of the meiosis in several species of *Aeshna* and observed that the bivalents orientated equationally during metaphase I. Cumming (1964a) described the course of meiosis in two individuals of *Orthemis levis* (Libellulidae) heterozygote for a fusion (2n = 5 = neo-XY) and observed that at metaphase I the fusion trivalent ("tripartite bivalent") was orientated equationally. All cells at meiosis II carried a "triple chromatid" as a result of the equational division of trivalent at metaphase II it adopted a "U" shape and divided reductionally at anaphase II.

The presence of heteromorphic sex chromosome at meiosis II was reported in *Aeshna coerula* and *Aeshna grandis* by Oksala (1943b) and Kiauta (1969a) respectively. Similar observations were made by Omura (1955) and Mola (1995) in two Libellulidae species *Pseudothemis zonata* and *Erythrodiplax media.* The heteromorphic autosomal bivalent of *Brachymesia furcata* (Libellulidae) was also observed at meiosis II. This could be taken as evidence, as far as equational division of heteromorphic bivalents at anaphase I in all these species (Agopian and Mola, 1988).

Kiauta (1975) admitted that in dragonflies, the structure of the bivalents at diakinesis or primary metaphase did not give any cue to the mode of meiosis with respect to division and distribution of the sex element. The only argument in favour of post-reduction was the observation that sex element divided mitotically at primary anaphase and preceded undivided to one pole or the other in second division. Nothing could be said, however, on the mode of reduction of the autosomal bivalents.

Mola (1995) explained and illustrated diagrammatically the behaviour of autosomal bivalents and sex bivalents and univalents in post and pre-reductional meiosis. In both types of meiosis, X chromosome was present at prometaphase II in the form of a chromosome with two chromatids of equal size, giving no information about the type of meiosis. However, the result would be different when heteromorphic bivalents were considered (such as sex bivalents). At post-reductional meiosis, these bivalents orientated at metaphase I with their long axis on the equational plane, while at prometaphase II, a chromosomes with two chromatids of unequal size migrated to each pole. At pre-reductional meiosis, heteromorphic bivalents orientated axially (with their long axis perpendicular to the equational plane) and at prometaphase II one cell received a larger chromosome (neo-X) with two equal sized chromatids, while the smaller chromosome (neo-X) migrated to the other pole. The sex univalents equationally orientated, divided equationally at anaphase I and all cells at mataphase II possessed 14 chromosomes.

Hughes-Schrader (1948) and Battalgia and Boyes (1955) demonstrated post-reductional meiosis in coccids (Homoptera) with holokinetic chromosomes. In these groups, at anaphase I, homologous chromatids migrated parallel to equational plane very close to one another or were even associated with their telomeric regions. In few Odonata species in which cells at anaphase I were observed (Oksala 1943b; Mola and Agopian 1985; Agopian and Mola 1988; Mola 1995), the migrating chromatids were joined by one telomeric region and they migrated almost parallel to the equatorial plane, this particular shape was retained and observed until prometaphase II stages.

Walia (1997) studied orientation of the elements, whether equational or axial, it was difficult to determine because of the small size of bivalents, similar morphology and close aggregation on the equatorial plate. She inferred post-reductional division of the X element based on the division and placement of the sex chromosome during two meiotic divisions. The sex element was divided mitotically during anaphase I, while during anaphase II whenever found possessed two groups of cells, one with X chromosome and other without X element. This depicted post-reductional division of the sex element in Odonata.

Cytologically, Odonata represent one of the most extensively studied order in all the insect groups. Yet the phylogenetic relationships within and between different taxa, nature and behaviour of centromere, origion of m chromosomes and mechanism of chromosome evolution in this group are not well understood. In recent years, Scientist apply more refined modern techniques in the field of Odonata cytology to solve various controversies regarding the karyotypic evolution, which will help in tracing phylogenetic relationships within and between different taxa of this group at cytotaxonomic level.

References

Agopian, S.S. and Mola, L.M., 1988. Intra and interspecific karyotype variability in five species of Libellulidae (Anisoptera, Odonata). *Caryologia*, 41(1): 69–78.

Asana, J.J. and Makino, S., 1935. A comparative study of the chromosomes in the Indian dragonflies. *J. Fac. Sci.*, Hokkaido, VI, 4(2): 67–86.

Battaglia, E. and Boyes, J.W., 1955. Post-reductional meiosis: Its mechanism and causes. *Caryologia*, 8: 87–134.

Boys, J.W., Vanbrink, J.M. and Kiauta, B., 1980. Sixteen dragonfly karyotypes from the Republic of South Africa and Swaziland, with evidence on the possible hybrid nature of *Orthetrum Jullia falsum* Longfield (Anisoptera: Libellulidae). *Odonatologica*, 9(2): 131–145.

Cruden, R.W., 1968. Chromosome number of some North American dragonflies (Odonata). *Can. J. Genet. Cytol.*, 10(1): 200–214.

Cumming, R.B., 1964a. Cytogenetic studies in the order Odonata. *Ph.D. Thesis*, University Texas, Austin.

Chatterjee, K. and Kiauta, B., 1973. Male germ cell chromosomes of two Calopterygoidea from the Darjeeling Himalaya (Zygoptera: Chlorocyhidae, Euphaeidae). *Odonatologica*, 2(2): 105–108.

Darlington, C.D., 1965. *Cytology*. Churchill, London.

Dasgupta, J., 1957. Cytological studies on the Indian dragonflies. II. A study of the chromosomes during meiosis in thirty species of Indian Odonata (Insecta). *Proc. Zool. Soc.*, Calcutta, 10(1): 1–66.

Ferreira, A., Kiauta, B. and Zaha, A., 1979. Male germ cell chromosomes of thirty two Brazilian dragonflies . *Odonatologica*, 8(1): 5–22.

Gama, V., Landim, C. da Cruz and Ferreirea, A., 1981. Estudos morfologicos da estrutura dos cinetocores de *Enallagma cheliferum* (Odonata). *Naturalia* (Sao Paulo), 6: 75–87.

Handa, S. M., Mittal, O.P. and Batra, H.N., 1984. Chromosomes in ten species of dragonflies (Anisoptera: Odonata). *Research Bulletin (Science)*, Punjab University, 35(3–4): 65–73.

Hogben, L., 1921. Studies on synapsis III. The nucleolar organization of the germ cell in *Libellula depressa. Proc. R. Soc.* Lond. (B), 92: 60–80.

Hughes-Schrader, S., 1948. Cytology of coccids (Coccidae–Homoptera). *Adv. Genet.*, 2: 127–203.

Kiauta, B., 1967e. Considerations on the evolution of the chromosome complement in Odonata. *Genetica*, 38: 430–446.

Kiauta, B., 1968b. Variation in size of the 'm'-chromosome in the dragonfly *Calopteryx virgo* (L) and its significance for the chorogeography and taxonomy of the *Calopteryx virgo* superspecies. *Genen Phaenen*, 12(1): 11–16.

Kiauta, B., 1968c. Variation in size of the dragonfly m-chromosome, with considerations on its significance for the chorogeography and taxonomy of the order Odonata, and notes on the validity of the rule of Reinig. *Genetica*, 39: 67–74.

Kiauta, B., 1968d. Morphology and kinetic behaviour of the odonate sex chromosomes, with a review of the distribution of sex determining mechanisms in the order. *Genen Phaenen.*, 12(2): 21–24.

Kiauta, B., 1969a. Sex chromosomes and sex determining mechanisms in Odonata, with a review of the cytological conditions in the family Gomphidae, and references to the karyotypic evolution in the order. *Genetica*, 40: 127–157

Kiauta, B., 1969b. Autosomal fragmentations and fusions in Odonata and their evolutionary implications. *Genetica*, 40: 158–180.

Kiauta, B., 1970a. Some remarks on the evolution of the centromere, based upon the distribution of centromere types in insects. *Ent. Ber. Amst.*, 30(4): 71–75.

Kiauta, B., 1971a. An unusual case of precocious segregation and chromosome fragmentation in the primary spermatocytes of the damselfly, *Calopteryx virgo meridionalis* Selys, 1873, as evidence for a possible hybrid character of some populations of the *Calopteryx virgo* complex (Odonata, Zygoptera, Caloptrygidae). *Genen Phaenen.*, 14(2).32–40.

Kiatua, B., 1975. Cytotaxonomy of dragonflies, with special reference to the Nepalese fauna. Nepal Research Center, Kathmandu, pp. 1–70.

Kiauta, B. and Kiauta, M.A.J.E., 1981. The karyotype of *Risiocnemis incisa* Kimmis., 1936, from Luxon, The Phillippines (Zygoptera: Platycnemidididae). *Odonatologia*, 10: 151–154.

Kiauta, B. and Kiauta, M.A.J.E., 1982. The chromosomes of sixteen dragonfly species from Arun Valley, Eastern Nepal. *Odonatologia,* 1(9): 143–145.

Kiauta, B. and Ochssee, G.A.B., 1979. Some dragonfly karyotypes from the Voltaic Republic (Haute Volta), West Africa. *Odonatologia,* 8(1): 47–54.

Lefevre, G. and McGill, C.C., 1908. The chromosomes of *Anasa tristis* and *Anax Junius. Amer. Jour. Anst.,* 7: 467–487.

Lima de Faria, A., 1949. Genetic origin and evolution of kinetochores. *Hereditas,* 35: 422–444.

Makalovskaja, W.N., 1940. Comparative karyological studies of dragonflies (Odonata). *Arch. Russ. D' anat., d'embryol,* Leningrad, 25(1): 24–39.

McGill, C., 1904. The spermatogenesis of *Anax junius.* Univ. MO. Stud. 2 (5): 236–250).

Mittal, O.P. and Gandhi, V., 1984. Chromosome make-up in two species of damselflies (Odonata: Zygoptera). Abstr. Pap. *Ist Indian Symp. Odonatol.,* Madurai, p. 6.

Mola, L.M., 1995. Post-reductional meiosis in *Aeshna* (Aeshnidae, Oconata). *Hereditas.* 122: 47–55.

Mola, L.M. and Agopian, S.S., 1985. Observations on the chromosomes of four species of South American Libellulidae (Anisoptera). *Odonatologica,* 14(2): 115–125.

Oguma, K., 1930. A comparative study of the spermatocyte chromosome in allied species of the dragonfly. *Jour. Fac. Sci.* I Iokkaido Imp. Univ. Scrv. VI, 1: 1–32.

Oksala, T., 1943b. Cytological studies on dragonflies, I. The chromosome conditions in the genus *Aeschna,* with special reference to the post-reductional division of bivalents. *Suom. tiedeakat. esit.* Poytak., pp. 196–197.

Oksala, T., 1944b. Sytologisia tutkimuksia Sudenkorennoillia, II. Meiotisen prekositeetin synnista (Cytological studies on dragonflies, II. Origin of the meiotic precocity). *Suom. tiedeakat. esit.* potok 1944: 198–199 (in Finish).

Oksala. T., 1945. Zytologische Studien an Odonaten III. Die Ovogense. *Ann. Acad. Sci. Fenn.,* (Ser. A IV), 9: 1–32.

Oksala, T., 1952. Chiasma formation and chiasma interference in the Odonata. *Hereditas*, 38: 449–480.

Omura, T., 1955. A comparative study of the spermatogenesis in the Japanese dragonflies I: Family Libellulidae. *Biol. J. Okoyama Univ.*, 2 (2–3): 95–135.

Perepelov, E. and Bugrov, A.G. and Warchalowska-Sliwa, E., 2001. C-band karyotypes, of some dragonfly species from Russia. II The families Cordulegasteridae, Corduliidae and Gomphidae. *Folia Biol.* (Krakow), 49(3–4): 175–178.

Piza de Toledo, S., 1950. The present status of the question of the kinetochore. *Genet. Iber.*, 2: 193–199.

Piza de Toledo, S., 1953. The crucial proffs of the dicentiricity of Hemiptera Chromosomes. *An Esc. Sup. agric.* 'Luis de Queiros' 10: 174–176.

Ray Chaudhury, S.P. and Dasgupta, J., 1949. Cytological studies on the Indian dragonflies I. Structure and behaviour of chromosomes in six species of dragonflies (Odonata). *Proc. Zool. Soc. Bengal*, 2(1): 81–93.

Saez, F.A., 1963. Gradient of the heterochromatization in the evolution of the sexual system 'N' 'neoX–neo Y'. *Partug. Acta. Biol.*, (A)4: 111–138.

Seshachar, B.R. and Bagga, S., 1962. Chromosome number of sex–determining mechanism in the dragonfly *Hemianax ephippiger* (Burmeister). *Cytologia*, 27: 443–449.

Smith, E.A., 1916. Spermatogensis of the dragonfly *Sympetrum semicinctum* (Say) with remarks. upon *Libellula basalis. Biol. Bull. Mar. Biol Lab.*, Woods Hole., 31: 269–290.

Srivastava, M.D.L. and Das. C.C., 1953. Heteropycnosis in autosome segments of *Ceriagrion coromadelianum* (Odonata). *Nature, Lond.* 172: 765.

Suzuki, K. and Saioth, K., 1988. Germ-Line chromosomes of two species of *Davidius* with special reference to the sex chromosomes (Anisoptera, Gomphidae). *Odonatologia*, 17(3): 275–280.

Thomas, K.I. and Prasad, R., 1984. Chromosomes and the chromosomal basis of sex determination in a few species of

Indian dragonflies (Odonata). In: *Abstr. Pap. Ist Indian Symp. Odonatol,* Madurai, p. 10.

Thomas, K.I. and Prasad, R., 1986. A study of the germinal chromosomes and C-band patterns in four Indian dragonflies (Odonata). *Perspectives in Cytology and Genetics,* 5: 125–131.

Tyagi, B.K., 1978b. Studies on the chromosomes of Odonata of Dun Valley (Dehradun, India). *Ph.D. Thesis,* University of Garhwal, Srinagar, (Garhwal).

Tyagi, B.K., 1982. Cytotaxonomy of the Indian dragonflies. *Indian Rev. Life. Sci.,* 2: 149–161.

Tyagi, B.K., 1986. Cytogenetics, Karyosystematics and Cytophylogeny of Indian Odonata. *Indian Rev., Life. Sci.,* 6: 221–239.

Walia, G.K., 1997. Cytological studies on some North and North–east Indian Odonata. *Ph.D. Thesis,* Punjabi University, Patiala, (Punjab).

Walia, G.K., 2006. C band pattern in four species of dragonflies. Abstr. Pap. *National Congress of Entomology,* Patiala, pp 145.

Walia, G.K. and Sandhu, R., 1999c. Autosomal fragmentations in five species of family Coenagrionidae (Zygoptera : Odonata). *Fraseria,* 5:15–20

White, M.J.D., 1948. *Cytogenetics.* Cambridge University Press, Cambridge.

White, M.J.D., 1957. Cytogenetics and systematic entomology. *Ann. Rev. Ent.,* 2: 71–90.

White, M.J.D., 1973. *Animal Cytology and Evolution,* 3rd edn. Cambridge University Press. Cambridge.

Chapter 16

Algal Contribution to Dissolved Oxygen in Small Hilly Streams: A Case Study

M.R. Sharma[1]* and A.B. Gupta[2]

[1]*Executive Engineer, Irrigation & Public Health Department, M-26, H.B. Colony, Hamirpur – 177 001, H.P*

[2]*Professor, Department of Civil Engineering, Malaviya N.I.T, Jaipur – 302 017, Rajasthan*

ABSTRACT

Water Quality models has gained wide acceptance as valuable tools to support the effective management of pollution-impacted streams and lakes. The credibility of the model is accomplished through model calibration and verification. Hathli is sub-tributary of river Beas in outer Himalayas. It is getting polluted due to wastewater of Hamirpur town in Himachal Pradesh, India. The pollution impact is severest in the low flow months. Steep slopes, pools, riffles and small waterfalls thus characterize the stream. The streambed consists of stones and cobbles. All the stones in the streambed are heavily coated with a greenish layer of algae. There are also plentiful greenish filamentous attached algae in the stream. Some aquatic plants have also been observed both in stream. The paper describes the

---

* *Corresponding Author.*

quantification of Oxygen production due to the presence of attached algae in the stream and need to modify the existing models to predict the water quality of hilly streams.

Keywords: *Water quality, Water quality modeling, Hathli stream.*

Introduction

Hathli is one of the sub tributaries of river Beas in Hamirpur district of Himachal Pradesh (India). It lies at latitude of 31°-25′ N and 76°–19′ East longitude. The Hamirpur town is located on Right Bank of a Hathli stream. It is a small rain fed perennial stream taking its orgin from near Townidevi and meandering over 10 km in the district of Hamirpur. It ultimately joins Kunah stream, which is a tributary of river Beas. It swells during rainy season but gets reduced to a narrow stream in the summer. The stream serves as drinking water source for the region. For want of proper sewerage system, the night soil from the houses is being treated through septic tanks. The water from kitchen and baths flows in open drains and is being discharged into local nallahs named as Manji nallah and Gaura nallah. The wastewater of the town is polluting the stream (Sharma *et al.*, 2002; Sharma *et al.*, 2003).

Water Quality models has gained wide acceptance as valuable tools to support the effective management of pollution-impacted streams and lakes. The credibility of the model is accomplished through model calibration and verification. The verified model can be used to forecast the water quality, when any control measure is implemented. During the water quality modeling of Hathli stream, it was observed that the water quality model Stream-1 assumes very high values of Deoxygenation Coefficient (K_1) and Reaeration coefficient (K_2) (Sharma and Gupta, 2004). The paper describes the quantification of Oxygen production due to the presence of attached algae in the stream. Thus there is a need to modify the existing water quality models to predict the water quality of hilly streams.

Stream Geometry

Hathli stream has a steep slope which varies greatly on different reaches. The stream is having different velocity and cross-section in different reaches. The stream comprises of small waterfalls and tiny pools. Steep slopes, pools, riffles and small waterfalls thus characterize the stream. The streambed consists of stones and cobbles. All the stones in the streambed are heavily coated with a greenish layer of algae. There are also plentiful greenish filamentous attached algae in the stream.

Materials and Methods

To assess the contribution of solar energy to DO, a section of Hathli stream, 400m in length *i.e.* Reach –1 (Figure 16.1) was selected and DO was monitored for one year, round the clock and diurnal variation in DO was observed.

Since the entire bed of reach-1 is covered with attached algae it has been assumed that the bed of stream is covered with a uniform, thin layer of algae growing on stones and rocks and the entire bed is contributing to the oxygen production. An attempt has been made to correlate diurnal DO variation with solar energy and algal growth.

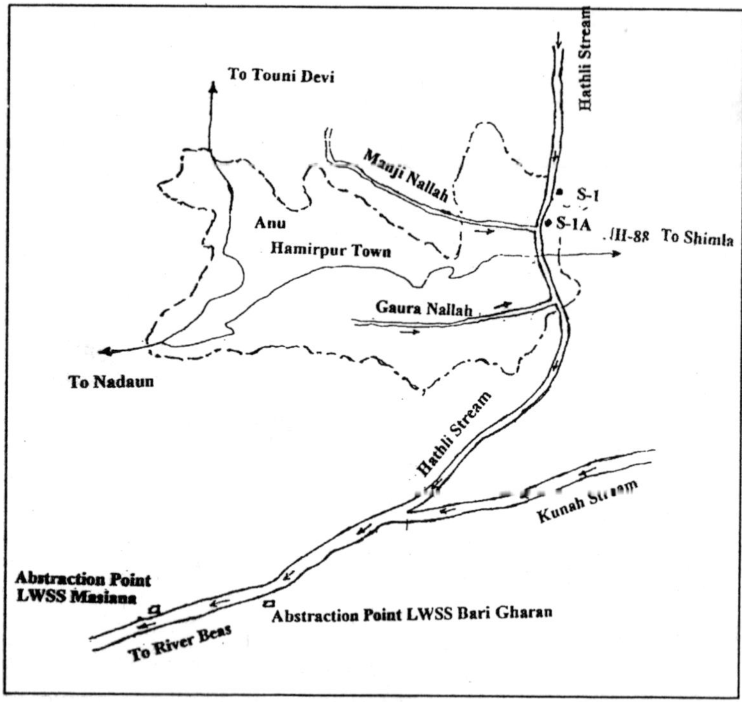

Figure 16.1: Location Plan of Sampling Station for Diurnal DO Measurement in Hathli Stream (Unpolluted Site)

Algal growth converts solar energy to chemical energy in the organic form. It has been assumed that 6 per cent of the visible light energy will get converted to algal growth in the Hathli stream. This is parallel to the assumption universally made for oxidation pond design. The chemical energy contained in an algal cell has been found taking average of 6000 calories per gram of algae, on ash free basis (Oswald, 1972).

The light energy received on a surface has been stated as gram calories per cm^2 per day (langleys/day). Visible radiation, which can penetrate the water surface, is limited to 4000-7000 A° (Arceivala, 1981). Typical values for latitudes from 12° to 34° North have been taken from standard tables. These ideal values have been adjusted for cloudiness and elevation. The "maximum" values relate to clear sky conditions, while "minimum" values occur when the sky is cloudy, depending on the "sky clearance factor" for the area. The Sky Clearance Factors (SKF) values for the Indian cities have been taken from standard tables (Bajwa, 1993). Since the SKF values for all the months and the site under study were not available, these values have been assumed very close to the values given for Roorkee city which is situated at about the same latitude and has similar weather conditions as Hamirpur. The possible day light hours for a particular month have been taken from standard tables (Michael, 1992). These values have also been compared with the local almanac available for the region. The average visible radiation received has been estimated as follows:

Average Radiation = Min. Radiation + [(Max. Radiation –
Min. radiation)] × Sky Clearance factor

Correction for Elevation above mean sea level

= Visible Radiation at sea level [1+ 0.003 EL]

where, EL is elevation in 100 m.

Applying this empirical formula average algal production in gram per square meter per day have been found for Reach-1 of the Hathli stream (Figure 16.1) *i.e.* 400m in length and oxygen production has been calculated by multiplying the algal production by a factor of 1.3. The oxygen consumed for algal respiration at the rate of 10 per cent of oxygen produced has been deducted from it. Average discharge of the stream in that stretch has been observed. The volume of the water passing that stretch has been calculated for the entire day light time. Average oxygen produced has been found out by multiplying the net oxygen produced with surface area of that reach. The value of oxygen available has been found out by dividing the average oxygen produced for the entire day light period by volume of water passing the bed area of stream during day light hours. This has been compared with the observed value of DO variation at site. The percentage deviation from the calculated values has been found out.

Keeping in view the fact that stream flow and concentration do not change rapidly, grab samples Dissolved Oxygen were collected at two stations from the center of stream at 0.6 depth. The guidelines given by USEPA in 'A Methods Manual for Volunteer Stream Monitoring, were followed for sampling (USEPA, 1997). All the samples were analysed following Standard Methods (APHA, 1992).

Results and Discussions

During the study it has been observed that the bed of Hathli stream is heavily coated with attached algae. In addition to it filamentous algae is also present in the stream water. The presence of rooted and attached macrophytes has also been noticed in Hathli stream as well as in Manji nallah. As a result of presence of attached algae in the water the diurnal variability of DO in the stream is high. The net effect of photosynthesis and respiration contributes to the average DO resources of the stream.

Hathli is a shallow stream and light can reach up to the bottom, the bottom plants (algae and macrophytes) would tend to make up most of the plant biomass. Heavy algal growths have been observed at the bed of stream growing on stones and rocks. Filamentous algae are also common at unpolluted and slightly polluted sites in the stream.

Table 16.1 gives the month-wise diurnal dissolved oxygen variation in Hathli stream at Reach–1. Table 16.2 gives the month-wise calculation of Oxygen production in Hathli stream at Reach–1, *i.e.* from RD 0 to 400 m.

Table 16.1: Diurnal Dissolved Oxygen (DO) Variation in Hathli Stream

Date	Day Time Hours															
	6.00	8.00	10.00	12.00	14.00	16.00	18.00	20.0	22.0	24.00	2.00	4.00	Mean	Max	Min	Max. Variation
	DO in mg/L															
Jan-02	8.9	9.4	11.1	11.9	11.0	10.8	10.6	10.4	9.8	8.8	8.8	8.8	10.0	11.9	8.8	3.1
Feb-02	7.7	8.9	10.6	11.6	11.1	9.6	9.4	9.2	7.7	7.7	7.7	7.7	9.1	11.6	7.7	3.9
Mar-02	7.1	8.3	10.1	11.0	10.6	9.1	8.9	8.7	7.1	7.1	7.1	7.1	8.5	11.0	7.1	3.9
Apr-02	6.8	7.3	7.8	8.9	9.9	8.7	7.3	6.4	6.4	6.4	6.4	6.4	7.4	9.9	6.4	3.5
May-02	5.2	6.5	7.3	8.2	8.8	10.8	9.2	7.7	6.2	5.2	5.2	5.2	7.1	10.8	5.2	5.6
Jun-02	5.3	7.1	7.8	9.2	9.8	8.9	7.5	5.3	5.3	5.3	5.3	5.3	6.8	9.8	5.3	4.5
Jul-02	5.1	5.3	7.0	7.0	6.5	6.5	5.9	5.6	5.4	5.3	5.3	5.3	5.9	7.0	5.1	1.9
Aug-02	5.4	6.2	7.6	7.6	7.2	6.8	6.4	5.9	5.4	5.4	5.4	5.4	6.2	7.6	5.4	2.2
Sep-02	6.0	7.5	9.0	9.3	9.0	8.0	7.6	6.0	6.0	6.0	6.0	6.0	7.2	9.3	6.0	3.3
Oct-02	6.3	7.5	8.5	8.8	7.5	6.3	6.0	6.0	6.0	6.0	6.0	6.0	6.7	8.8	6.0	2.8
Nov-02	8.0	8.8	9.5	10.5	10	8.4	8.3	7.9	7.9	7.9	7.9	7.9	8.6	10.5	7.9	2.6
Dec-02	8.5	9.2	10.2	10.8	10.8	9.5	8.6	8.5	8.5	8.5	8.5	8.5	9.2	10.8	8.5	2.3

Table 16.2: Solar Radiation and Algal Production/Oxygen Production in Hathli Stream in Reach–1

Month	*Visible Radiation*		*Sky Clearance Factor*	*Average Radiation*	*Correction for Elevation for 700 m*	*Average Algal Production, kg/ha-day*	*Average Algal Production, sqm-day*	*Average Oxygen Produced, sqm-day*	*Oxygen Consumed for Algal Respiration at the Rate of 0.025*
	Max	*Min*							
1	*2*	*3*	*4*	*5*	*6*	*7*	*8*	*9*	*10*
Jan	126	63	60	100.8	102.95	102.95	10.30	13.38	0.33
Feb	169	87	60	136.2	139.11	139.11	13.91	18.08	0.45
March	212	126	60	177.6	181.40	181.40	18.14	23.58	0.59
April	258	146	70	224.40	229.20	229.20	22.92	29.80	0.74
May	290	181	80	268.2	273.93	273.93	27.39	35.61	0.89
June	296	166	65	250.50	255.85	255.85	25.59	33.26	0.83
July	289	178	65	250.15	255.50	255.50	25.55	33.21	0.83
Aug	269	163	65	231.90	236.86	236.86	23.69	30.79	0.77
Sept	226	140	80	208.80	213.26	213.26	21.33	27.72	0.69
Oct	185	104	90	176.90	180.68	180.68	18.07	23.49	0.59
Nov	138	80	85	129.30	132.06	132.06	13.21	17.17	0.43
Dec	114	60	85	105.90	108.16	108.16	10.82	14.02	0.35

Contd...

Table 16.2–Contd...

Net Oxygen Available mg/m² day	Average Day Light Hours during the Month at Hamirpur	Average Discharge in l.p.s. During the Month in Reach–1	Volume of Water Passing Bed Area During the Day Light Hours of Stream in Reach–1	Surface area of the Reach Producing Algal Oxygen	Average Oxygen Produced and Available for Mass Transfer in Reach–1 of the Stream	Calculated Value of Oxygen, mg/L	Observed Value of Oxygen mg/L
11	12	13	14	15	16	17	18
13.05	10.30	76	2818080	534	6968403	2.5	3.1
17.63	11.00	77	3049200	528	9309847	3.1	3.9
22.99	11.90	60	2570400	428	9840524	3.8	3.9
29.05	12.80	30	1382400	314	9121872	6.6	3.5
34.72	13.60	10	489600	176	6110868	12.5	5.6
32.43	14.00	30	1512000	232	7523626	5.0	4.5
32.38	13.81	350	17400600	982	31801198	1.8	1.9
30.02	13.15	500	23670000	1350	40529011	1.7	2.2
27.03	12.25	250	11025000	900	24327893	2.2	3.3
22.90	11.15	200	8028000	850	19466069	2.4	2.8
16.74	10.50	123	4649400	615	10294497	2.2	2.6
13.71	10.10	85	3090600	543	7444359	2.4	2.3

Illustration

The sample calculations for the month of January are given below:

1. Visible Solar Radiation for the month of January

   Maximum = 126

   Minimum = 63

2. Sky Clearance factor (S.K.F) = 60 per cent

3. Average Radiation = Min. Rad. (Max. Radiation. – Min Radiation) × S.K.F

   = 63 + (126–63) 0.60 = 100.8

4. Applying Correction for Elevation for 700 m above mean sea level

   = Visible Radiation at sea level [1+ 0.003 EL],

   where, EL is elevation in 100 m

   = 100.8[1 + 0.003 × 7] = 102.95

5. Average algal production in kg/ha – day

   = {100.8 x 10^8 x 0.06} × (1/6000) × (1/10^3)

   = 102.95 kg/ha – day

6. Average algal production in g/m^2 – day

   = 102.95 × 1000 x (1/10000)

   = 10.30 g/Sq.m

7. Average oxygen production at the rate of 1.3 times the algal production

   = 10.30 × 1.3 = 13.38 g/Sq.m

8. Oxygen consumed for algal respiration

   = 0.025 × 13.38 = 0.33 g/sq.m

9. Net oxygen available in g/sq.m per day

   = 13.38 – 0.33 = 13.05 g/Sq.m

10. Average day light hours during the month of January at Hamirpur

    = 10.30 hrs.

11. Average discharge in liter per second during the month at Reach–1

    = 76.0 l.p.s

12. Volume of water passing the bed area of stream during the day light hours at Reach–1

    = 2818080 litres

13. Surface area of the Reach contributing to algal/oxygen production

    = 534 Sq.m

14. Average oxygen produced in mg and available for mass transfer in Reach–1 of stream

    = (Net oxygen available in g/Sq.m) × 1000 × (Surface area of Reach–1)

    = 13.05 × 1000 × 534 = 6968700 mg

15. Calculated value of oxygen available in mg/l

    = (Average oxygen produced in mg and available for mass transfer in Reach -1 of stream) ÷ (Net oxygen available in g/Sq.m × 1000 × Surface area of Reach–1)

    = 6968700 ÷ 2818080 = 2.5 mg/L

16. Observed Value of D.O variation in mg/l at Reach–1 in stream

    = 3.1 mg/l

17. Perc ent deviation from the calculated value

    = (0.6/2.5) × 100 = 24 per cent

It can be observed from Table 16.2, that with the exception of values for the month of April, May and September, the observed values of Oxygen variation range from 90 per cent to 130 per cent of the calculated values. For the month of April and May the observed values are only 53 per cent and 45 per cent of the calculated values respectively. It is due to the reason that oxygen level during day, in the month of April and May reaches a peak value of 9.9 and 10.8 mg/l respectively, which is much higher than the saturation values of oxygen (Table 16.4) at that particular temperature. Since the stream is flowing the dissolved oxygen may be escaping to the atmosphere.

In general, values of the average daily, areal production rate range from about 0.3 to 3 g O m^{-2} d^{-1} for moderately productive streams. Highly productive systems can range from 3 to 20 gm Oxygen per square meter per day (Chopra, 1997). Thus it can be concluded that the algal biomass is major contributor of oxygen in the stream and must be taken into account during water quality modeling of the stream.

Table 16.3: Seasonal Average of Diurnal DO Variation in Hathli Stream at Un-polluted Site

Sl.No.	*Season*	*Month*	*Observed Value of Diurnal Variation of DO in the Stream in mg/l*	*Remarks*
1.	Winter	October	2.8	Observed values
2.	"	November	2.6	have been
3.	"	December	2.3	taken from
4.	"	January	3.1	Table 16.2
5.	"	February	3.9	
		Min. Value of Season	2.3	
1.	Summer	March	3.9	
2.	"	April	3.5	
3.	"	May	5.6	
4.	"	June	4.5	
5.	"	Min. Value of Season	3.5	
1.	Monsoon	July	1.9	
2.	"	August	2.2	
3.	"	September	3.3	
		Min. Value of Season	1.9	

Conclusions

The streams in hilly regions of outer Himalayas are generally small, having shallow with steep and variable slope. The streambeds of these rivulets are covered with attached algae. The floating algae are generally insignificant due to swift and shallow nature of streams. These streams have high diurnal as well as seasonal variation in DO levels. Though QUAL-2e water quality model takes into account the effect of floating algae in rivers but the effect of attached algae has not been considered in any of the known models. Since the effect of attached algae on DO levels in shallow streams are quite

Table 16.4: Max. Dissolved Oxygen at Various Temperature at Sea Level and at a Height of 700 m Above Mean Sea Level *i.e.* Mean Level of Hathli Stream

Temperature in Degrees Celsius	*D.O at Sea Level (mg/l)*	*D.O at a Height of 700m Above Mean Sea Level (mg/l)**	*Temperature in Degrees Celsius*	*D.O at Sea Level (mg/l)*	*D.O at a Height of 700m Above Mean Sea Level (mg/l)**
0	14.6	13.42	23	8.56	7.87
1	14.19	13.04	24	8.40	7.72
2	13.81	12.70	25	8.24	7.58
3	13.44	12.36	26	8.09	7.43
4	13.09	12.04	27	7.95	7.31
5	12.75	11.73	28	7.81	7.18
6	12.43	11.43	29	7.67	7.05
7	12.12	11.14	30	7.54	6.93
8	11.83	10.88	31	7.41	6.81
9	11.55	10.62	32	7.28	6.69
10	11.27	10.36	33	7.16	6.58
11	11.01	10.12	34	7.05	6.48
12	10.76	9.89	35	6.93	6.37

Contd...

Table 16.4–Contd...

Temperature in Degrees Celsius	*D.O at Sea Level (mg/l)*	*D.O at a Height of 700m Above Mean Sea Level (mg/l)**	*Temperature in Degrees Celsius*	*D.O at Sea Level (mg/l)*	*D.O at a Height of 700m Above Mean Sea Level (mg/l)**
13	10.52	9.67	36	6.82	6.27
14	10.29	9.46	37	6.71	6.17
15	10.07	9.26	38	6.61	6.07
16	9.85	9.06	39	6.51	5.99
17	9.65	8.87	40	6.41	5.89
18	9.45	8.69	41	6.31	5.80
19	9.26	8.51	42	6.22	5.72
20	9.07	8.34	43	6.13	5.64
21	8.90	8.18	44	6.04	5.55
22	8.72	8.02	45	5.95	5.47

Source: Volunteer Stream Monitoring–A Methods Manual, (1997) EPA 841-B-97-003, p.140

Note: * Saturation Concentration of D.O at 'p' Atmospheric pressure = Saturation Conc of DO at 1 atm {1- 0.1148 x elev. (km)}

Where, elev is Elevation above mean sea level (Zison 1978).

significant, there is need to modify the models to take into account the presence of periphyton i.e. benthic algae.

Acknowledgement

The authors are grateful to Er. R.N. Sharma, Engineer-in-Chief, Er. P.S. Verma, Er. O.P. Chauhan XENs of I and P.H Deptt., and Sh. Vishnu Prashad Lab Attendant, for their help during the study.

References

American Public Health Association (APHA), 1992. *Standard Methods for the Examination of Water and Wastewater*, 18th edn. Washington, DC, p. 874.

Arceivala, S., 1981. *Wastewater Treatment and Disposal*. Marcel Dekker, New York.

Bajwa, G.S., 1993. *Practical Handbook on Public Health Engineering*. Deep Publishers, Shimla.

Chopra, S.C., 1997. *Surface Water Quality Modeling*. McGraw Hill, N.Y.

Michael, A.M., 1992. *Irrigation: Theory and Practice*. Vikas Publishing House, New Delhi.

Oswald, W.J., 1972. Complete waste treatment in ponds. In: *Proceedings of 6th International Water Pollution Research Conference*, Pergamon Press, London.

Sharma, M.R. and Gupta, A.B., 2004. Prevention and control of pollution in streams of outer Himalayas: A case study. In: *36th Annual Convention of IWWA*, Ahmedabad, January 9–13, p. 77–85.

Sharma, M.R., Gupta, A.B. and Bassin, J.K., 2002. Water quality management of Hathli stream in lower Himalayas. *Indian Journal of Environmental Protection*, 9: 1014–1025.

Sharma, Moti Ram, Bassin, J.K. and Gupta, A.B., 2003. A pollutional profile of Hathli stream in lower Himalayas. *Journal of Pollution Research*, 22(2): 237–240.

Qual-2E UNCAS, 1987. *Documentation and User Model*, USEPA, Athens.

USEPA, 1997. *Volunteer Stream Monitoring: A Methods Manual*. Office of Water, 4503F, EPA 841–B–97–003.

Chapter 17

Aquatic Environment of Ram Ganga River at Moradabad, India: A Quantitative Assessment

D.K. Sinha[1], Shilpi Saxena[2]*

[1]Reader, [2]Research Scholar,

K.G.K.(P.G.) College, Department of Chemistry, Moradabad – 244001

ABSTRACT

Water Quality Index (WQI) for Ram Ganga river water at Moradabad for twelve different sites for pre-monsoon period as well as after the onset of monsoon has been calculated with the help of W.H.O. water quality standards and the data of fifteen water quality physico-chemical parameters estimated following standard methods and procedures. River water at almost all the sites is found to be severely polluted during the course of study. To study the pollution potential of effluents and effect of different kinds of human activities, the river water quality at two different sites is also assessed on the basis of water quality indices of downstream river water samples after the mixing up of two different effluents carrying mixed discharge of industries and nearby locality. In general, to some extent the river

* *Corresponding Author.*

water quality shows improvement after the onset of monsoon. The present study is a standing example of man's disregard to the nature.

Keywords: *Water Quality Index (WQI), Physico-chemical parameters, Assigned unit weight (W_n), Quality rating (q_n).*

Introduction

Man's interest in water is as old as the history of man himself on earth. Life is supposed to have originated in water and water is the most essential requirement of all lives. The human cultures evolved along the river courses and man had used this resource for disposal of his wastes, irrigation of his crops, and more recently for industrial development. All these activities make them a subject of detailed scientific investigation.

"All ancient civilizations emerged on the banks of rivers. Rivers have been central to the growth of the human society. The river is an indicator of standard of the society. A dirty river means dirty society."

Pollution of environment is one of the most horrible ecological crisis to which we are subjected today. Sometimes in the past, three basic amenities–air, land or soil and water were pure, virgin, undisturbed, uncontaminated and basically most hospitable for living organisms. But, the situation is reverse now and probably this is because of urban-industrial technological revolution and speedy exploitation of every bit of natural resources. Time is perhaps not too far when pure and clean water particularly is densely populated, industrialized water scarce areas may be inadequate for maintaining the normal living standards.

Many rivers of the world receive heavy flux of sewage, industrial effluents, domestic and agricultural waste which consists of substances varying from simple nutrients to highly toxic hazardous chemicals. In short, it can be said that water resources all over the world are threatened not only by over exploitation and poor management but also by ecological degradation.

Moradabad is a 'B' class city of Uttar Pradesh having urban population more than 41 lac. Moradabad is situated at the bank of Ram Ganga river and its altitude from sea level is about 670 feet. It is extended from Himalaya in North to Chambal river in South.

Moradabad is at 28°20′, 29°15′N and 78°4′, 79°E. District Bijnor and Nainital are in the North, Rampur is in the East, Ganga river is in the West and district Budaun is in the South of district Moradabad. Moradabad has seen rapid industrialization during last few decades although Brassware industry is an age old for Moradabad. The other major industries are Steelware, Paper mills, Sugar mills, Crushers, Dye factories and a number of ancillaries and small-scale industries related to these industries etc. Most of these industries are dumping their effluents in the two major rivers of Moradabad-Ram Ganga, Gagan river and thereby polluting it severely. City discharge and different kinds of human activities are also causing river water pollution.

The area is characterized by periodic occurrence of hot summers, moderate rains and cold winters. The maximum and minimum atmospheric temperatures are 48°C and 3°C respectively. The average rainfall varies between 800 to 1000 mm. The relative humidity is up to 90 per cent in monsoon season and in drier part of the year it decreases to less than 20 per cent .

Ram Ganga river is a tributary of very important and holy river of the country–The Ganga. It emerges from lower region of Shivalik particularly from a small place Lamba of district Chamoli of Uttaranchal at the height of 3069 meter. Moradabad is the first major city in the way of flow of Ram Ganga river. A dam for irrigational and power generation purposes is made on this river at Kalagarh, Uttaranchal. The Ram Ganga river mixes up with Ganga river at its left bank at Sherpur village of district Farrukhabad, Uttar Pradesh where another river Kali also mixes up with it. Ram Ganga river is a perennial river. Ram Ganga river is used for bathing, drinking, cremation, dumping of carcases and sewage, other human activities in addition to industrial discharge. Due to all these activities the Ram Ganga has become a health hazard for the people on the banks of the river and also for the people using directly or indirectly the river water. The pollution of Ram Ganga river water is also adversely affecting the flora and fauna of aquatic environment which is an essential part of the aquatic eco-system. Ram Ganga river has a total drainage area of about 500 km^2. The total stretch of river can be divided into three regions – (*a*) The upstream area; a fertile region, (*b*) The mid region; full of sand, (*c*) The down stream region; most alluvial land which is fertile.

Water Quality Index (WQI) has been regarded as one of the most effective way to communicate water quality. In a number of nation-wide studies, it has been found to be quite useful in the assessment of drinking water as well as river water quality. (Chaudhari 2004, Pradhan 2001, Singh 1999, Sinha 1989) Water quality of major Indian rivers had been assessed very effectively with the help of calculation of Water Quality Indices with the data obtained through quantitative analysis. (Sinha 2004, Sinha 1994)

Materials and Methods

Twelve different sites at district Moradabad were selected in order to study the physico-chemical characteristics of Ram Ganga river water samples for pre-monsoon period and after the onset of monsoon. Out of these twelve sites, ten samples are of Ram Ganga river water and the rest two are effluent samples carrying mixed discharge of industries and nearby locality.

In order to calculate Water Quality Index of water samples, fifteen water quality physico-chemical parameters were estimated quantitatively following standard methods and procedures using all the chemicals of anal R grade. (APHA 1995) The estimated physico-chemical parameters are–pH value, turbidity, conductivity, alkalinity, total solids (T.S.), total dissolved solids (T.D.S.), chloride, hardness, calcium, magnesium, free CO_2, dissolved oxygen (D.O.), biological oxygen demand (B.O.D.), chemical oxygen demand (C.O.D.) and fluoride.

A brief description of sampling sites is given below:

M.D.A. Colony

This is Ram Ganga river water at Moradabad development authority colony which is about 5 km. North to the Moradabad city. This is the site from where river enters in the city. Sand digging and occasional bathing activities are noticed at this site.

Jigar Vihar Colony

This is river water sample at Jigar Vihar colony. It is about 2 km. South to the site I and its water is used for washing, laundering and bathing activities. Cattle bathing is a regular feature at this site.

Kaliji Temple

Ram Ganga river water sample at very famous religious Kaliji Temple. It is about 2 km. far from the previous site. Being adjacent to the holy temple, bathing activities are common at religious festivals. Some cremation activities are also noticed at the left bank of the river at this site.

Upstream Nawabpura

This is Ram Ganga river water sample at densely populated Nawabpura Mohalla before mixing up of Nawabpura nullah. It is about 500 meter away to the Site no. III.

Nawabpura Nullah

This is a city discharge coming through a nullah of bigger width of about 3 meter. The discharge carries the mixture of industrial effluents and city discharge. The sample is having an objectionable odour. The odour is particularly because of hydrogen sulphide.

Downstream Nawabpura

This is Ram Ganga river water sample at Nawabpura after the mixing up of Nawabpura nullah with it. The sample is collected at a distance of about 50 meter from the point of mixing of nullah water with river water.

Upstream Barbalan

This is river water sample at mohallah Barbalan before mixing up of Barbalan nullah. This site is about 2 km away from Kaliji temple. Some bathing and cattle activities are noticed at this site at the time of sampling.

Barbalan Nullah

This is discharge sample at Barbalan carrying mixed discharge of Dhobi – Ghat, small-scale industries and nearby locality.

Downstream Barbalan

This is Ram Ganga river water sample collected at a distance of about 50 meter after the mixing up of Barbalan nullah with Ram Ganga river water. No relevant activity of our concern is noticed during the course of study.

Moradabad–Bareilly Railway Bridge

This is river water collected at Moradabad – Bareilly railway bridge. It is about 3 km South to Kaliji Temple. Fishing and cattle bathing are regular activities at this site.

Moradabad–Bareilly Roadways Bridge

This is river water collected at Moradabad – Bareilly roadways bridge which is about 400 meters away from Site no. X. Some agricultural fields are at the left bank of the river.

Lucknow–Delhi Bypass Bridge

This is Ram Ganga river water sample at Lucknow – Delhi bypass bridge which is about 1.5 km away from roadways bridge. Agricultural activities are also noticed at the left bank of the river. This is the site from where the Ram Ganga river leaves the Moradabad city. This site is selected to assess the river water quality at the time of leaving the Moradabad city.

The map of sampling sites can be viewed in Figure 17.1.

Water Quality Indices have been calculated with the help of the method proposed by Tiwari *et al.* (Tiwari 1986, Tiwari 1985). In this method following equations have been used:

$$\text{Quality rating, } q_n = 100[(V_n - V_{i)}/(V_s - V_i)]$$

where,

V_n: Actual amount of nth parameter
V_i: The ideal value of this parameter
V_i: 0 except for pH and D.O.
V_i: 7.0 mg/lit for pH
V_i: 14.6 mg/lit for D.O.
V_s: Its standard.

Unit weight (W_n) for various parameters is inversely proportional to the recommended standard ($S_{n)}$ for the corresponding parameter.

$$W_n = K/S_n$$

$\Sigma W_n = 1$, considered here

$$\text{Sub indices, } (SI)_n = (q_n)^{Wn}$$

The overall WQI is calculated by taking the geometric mean of these sub indices.

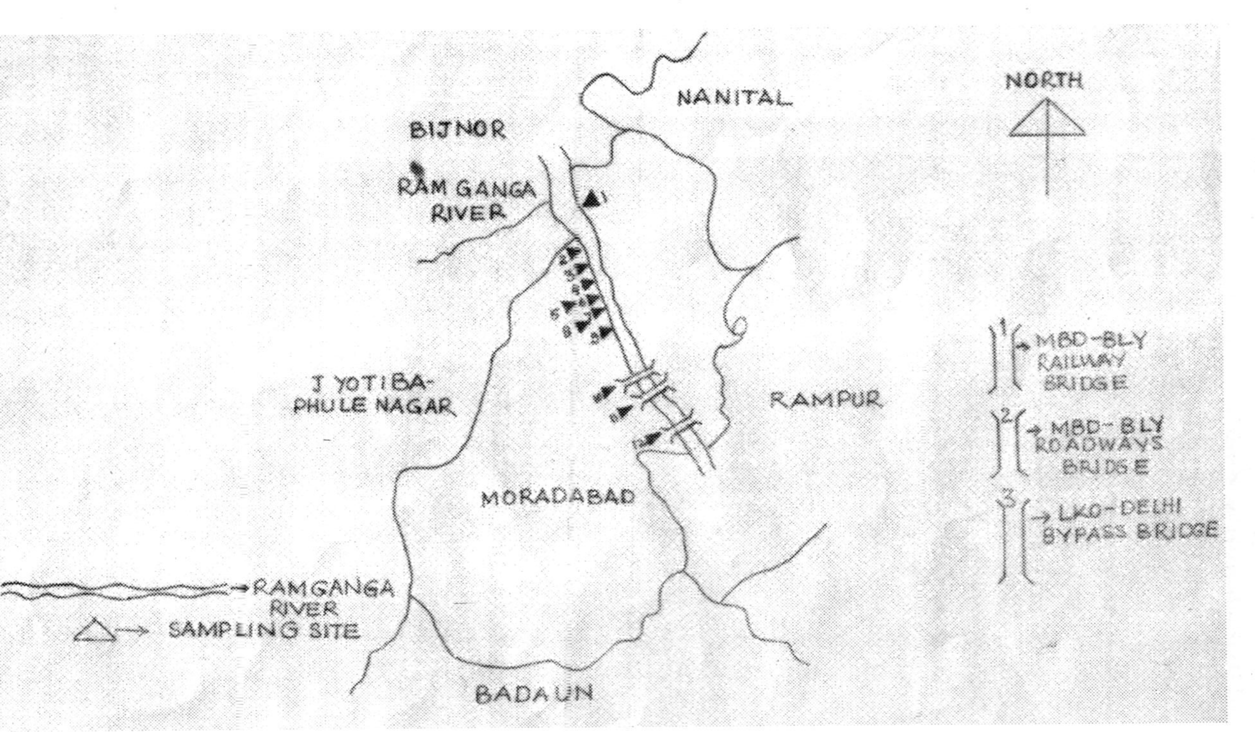

Figure 17.1: Map of Sampling Sites

$$WQI = \Sigma(SI)_n = \Sigma(q_n)^{W_n}$$

or

$$WQI = \text{antilog}_{10}[\Sigma W_n \log_{10} q_n]$$

For Water Quality Index of water samples, following assumptions have been made to assess the extent of pollution. This assumption has been tested in many ways and is found to be correct and very fruitful in assessing the water quality. The assumptions are

WQI < 50 – fit for human consumption

WQI < 80 – moderately polluted

WQI > 80 – excessively polluted

WQI >100 – Severely polluted

The Water Quality Index of twelve different sites at Moradabad district for pre-monsoon period and after the onset of monsoon are calculated with the help of data of fifteen water quality physico-chemical parameters using W.H.O. standards.

The purpose of calculating WQI and comparing it with the standards is to assess the Ram Ganga river water quality at Moradabad. To assess the pollution potential of effluents and different kinds of human activities, the river water quality is also assessed after the mixing up of effluents. The present study is also aimed to study the variation of river water quality after the onset of monsoon on the basis of calculated Water Quality Index.

Results and Discussion

Physico-chemical parameters, their W.H.O. standards and assigned unit weights of different parameters are listed in Table 17.1. Site-wise estimated actual value (V_n), calculated quality rating (q_n) and $W_n \log_{10} q_n$ for different parameters for pre-monsoon period are given in Tables 17.2 to 17.13. Likewise the values of (V_n), (q_n) and $W_n \log_{10} q_n$ for different water quality parameters after the onset of monsoon are presented in Tables 17.14 to 17.25. The finally calculated site-wise values of Water Quality Index for pre-monsoon period and after the onset of monsoon are enlisted in Table 17.26. Site-wise seasonal variation of Water Quality Index is presented graphically in Figure 17.2.

Table 17.1: Physico-chemical Parameters, Their W.H.O. Standards and Assigned Unit Weights

Sl.No.	Parameter	W.H.O. Standard	Assigned Unit Weight (W_n)
1.	pH	8.0	0.023656320
2.	Turbidity (NTU)	5.0	0.037850113
3.	Conductivity (μS/cm)	0.300	0.630835223
4.	Alkalinity (mg/lit)	100.0	0.001892505
5.	T.S. (mg/lit)	500.0	0.000378501
6.	T.D.S. (mg/lit)	500.0	0.000378501
7.	Chloride (mg/lit)	200.0	0.000946252
8.	Hardness (mg/lit)	100.0	0.001892505
9.	Calcium (mg/lit)	100.0	0.001892505
10.	Magnesium (mg/lit)	30.0	0.006308352
11.	Free CO_2 (mg/lit)	10.0	0.018925056
12.	D.O. (mg/lit)	5.0	0.037850113
13.	B.O.D. (mg/lit)	6.0	0.031541761
14.	C.O.D. (mg/lit)	10.0	0.018925056
15.	Fluoride (mg/lit)	1.0	0.189250567

Table 17.2: Parameter-wise Actual Value (V_n), Quality Rating (q_n) and Calculated Value of $W_n \log_{10} q_n$ at Site No. I for Pre-monsoon Period

Sl.No.	Parameter	Actual Value (V_n)	Quality rating (q_n)	$W_n \log_{10} q_n$
1.	pH	7.59	59.0	0.041891
2.	Turbidity (NTU)	2.48	49.6	0.064174
3.	Conductivity (μS/cm)	0.454	151.0	1.374575
4.	Alkalinity (mg/lit)	120.0	120.0	0.003934
5.	T.S. (mg/lit)	138.0	27.6	0.000545
6.	T.D.S. (mg/lit)	104.0	20.8	0.000498
7.	Chloride (mg/lit)	39.76	19.88	0.001228
8.	Hardness (mg/lit)	136.0	136.0	0.004037
9.	Calcium (mg/lit)	19.24	19.24	0.002430
10.	Magnesium (mg/lit)	124.28	414.26	0.016510
11.	Free CO_2 (mg/lit)	83.6	836.0	0.055302
12.	D.O. (mg/lit)	3.24	118.3	0.078462
13.	B.O.D. (mg/lit)	35.64	594.0	0.086535
14.	C.O.D. (mg/lit)	24.8	248.0	0.045315
15.	Fluoride (mg/lit)	0.203	20.3	0.247444

Table 17.3: Parameter-wise Actual Value (V_n), Quality Rating (q_n) and Calculated Value of $W_n \log_{10} q_n$ at Site No. II for Pre-monsoon Period

Sl.No.	Parameter	Actual Value (V_n)	Quality rating (q_n)	$W_n log_{10} q_n$
1.	pH	7.80	80.0	0.45020
2.	Turbidity (NTU)	2.84	56.8	0.066402
3.	Conductivity (µS/cm)	0.456	152.0	1.376358
4.	Alkalinity (mg/lit)	110.0	110.0	0.003863
5.	T.S. (mg/lit)	282.0	56.4	0.000626
6.	T.D.S. (mg/lit)	226.0	45.2	0.001256
7.	Chloride (mg/lit)	42.6	21.3	0.004013
8.	Hardness (mg/lit)	132.0	132.0	0.002496
9.	Calcium (mg/lit)	20.84	20.84	0.016398
10.	Magnesium (mg/lit)	119.30	397.67	0.051861
11.	Free CO_2 (mg/lit)	55.0	550.0	0.079155
12.	D.O. (mg/lit)	2.75	129.39	0.079155
13.	B.O.D. (mg/lit)	16.2	270.0	0.076689
14.	C.O.D. (mg/lit)	52.0	520.0	0.051400
15.	Fluoride (mg/lit)	0.245	24.5	0.262900

Table 17.4: Parameter-wise Actual Value (V_n), Quality Rating (q_n) and Calculated Value of $W_n \log_{10} q_n$ at Site No. III for Pre-monsoon Period

Sl.No.	Parameter	Actual Value (V_n)	Quality rating (q_n)	$W_n log_{10} q_n$
1.	pH	7.26	26	0.033473
2.	Turbidity (NTU)	2.64	52.8	0.065201
3.	Conductivity (µS/cm)	0.457	152.0	1.376383
4.	Alkalinity (mg/lit)	105.0	105.0	0.003825
5.	T.S. (mg/lit)	498.0	99.6	0.000756
6.	T.D.S. (mg/lit)	42.4	84.8	0.000729
7.	Chloride (mg/lit)	44.02	22.01	0.001270
8.	Hardness (mg/lit)	160.0	160.0	0.004171
9.	Calcium (mg/lit)	27.25	27.25	0.002716
10.	Magnesium (mg/lit)	143.39	477.98	0.016902
11.	Free CO_2 (mg/lit)	52.8	528.0	0.051525
12.	D.O. (mg/lit)	2.92	121.70	0.078928
13.	B.O.D. (mg/lit)	16.2	270.0	0.076689
14.	C.O.D. (mg/lit)	75.2	752.0	0.054432
15.	Fluoride (mg/lit)	0.203	20.3	0.247444

Table 17.5: Parameter-wise Actual Value (V_n), Quality Rating (q_n) and Calculated Value of $W_n log_{10} q_n$ at Site No. IV for Pre-monsoon Period

Sl.No.	Parameter	Actual Value (V_n)	Quality rating (q_n)	$W_n log_{10} q_n$
1.	pH	6.49	51.0	0.040394
2.	Turbidity (NTU)	2.76	55.2	0.065932
3.	Conductivity (μS/cm)	0.549	183.0	1.427233
4.	Alkalinity (mg/lit)	180.0	180.0	0.004268
5.	T.S. (mg/lit)	1902.0	380.4	0.000976
6.	T.D.S. (mg/lit)	472.0	94.4	0.000747
7.	Chloride (mg/lit)	51.12	25.56	0.001331
8.	Hardness (mg/lit)	156.0	156.0	0.004150
9.	Calcium (mg/lit)	40.88	40.88	0.003049
10.	Magnesium (mg/lit)	131.09	436.97	0.016656
11.	Free CO_2 (mg/lit)	74.8	748.0	0.54388
12.	D.O. (mg/lit)	5.67	93.02	0.074510
13.	B.O.D. (mg/lit)	48.6	810.0	0.091738
14.	C.O.D. (mg/lit)	46.4	464.0	0.031538
15.	Fluoride (mg/lit)	0.207	20.7	0.049048

Table 17.6: Parameter-wise Actual Value (V_n), Quality Rating (q_n) and Calculated Value of $W_n log_{10} q_n$ at Site No. V for Pre-monsoon Period

Sl.No.	Parameter	Actual Value (V_n)	Quality rating (q_n)	$W_n log_{10} q_n$
1.	pH	6.40	60.0	0.42064
2.	Turbidity (NTU)	3.52	70.4	0.069930
3.	Conductivity (μS/cm)	1.93	643.0	1.771518
4.	Alkalinity (mg/lit)	630.0	630.0	0.005297
5.	T.S. (mg/lit)	2800.0	560.0	0.001040
6.	T.D.S. (mg/lit)	1628	325.6	0.009510
7.	Chloride (mg/lit)	328.02	164.01	0.002095
8.	Hardness (mg/lit)	388.0	388.0	0.004899
9.	Calcium (mg/lit)	177.96	177.96	0.004258
10.	Magnesium (mg/lit)	279.57	931.92	0.018731
11.	Free CO_2 (mg/lit)	286.0	2860.0	0.065411
12.	D.O. (mg/lit)	9.72	50.83	0.064576
13.	B.O.D. (mg/lit)	119.88	1998.0	0.104106
14.	C.O.D. (mg/lit)	52.8	528.0	0.051525
15.	Fluoride (mg/lit)	0.252	25.2	0.265215

Table 17.7: Parameter-wise Actual Value (V_n), Quality Rating (q_n) and Calculated Value of $W_n log_{10} q_n$ at Site No. VI for Pre-monsoon Period

Sl.No.	*Parameter*	*Actual Value (V_n)*	*Quality rating (q_n)*	*$W_n log_{10} q_n$*
1.	pH	7.21	21	0.031278
2.	Turbidity (NTU)	2.64	52.8	0.065201
3.	Conductivity (µS/cm)	1.96	653.0	1.775746
4.	Alkalinity (mg/lit)	460.0	460.0	0.005039
5.	T.S. (mg/lit)	2530.0	506.0	0.001023
6.	T.D.S. (mg/lit)	2344.0	468.8	0.001010
7.	Chloride (mg/lit)	266.96	133.48	0.002011
8.	Hardness (mg/lit)	346.0	346.0	0.004805
9.	Calcium (mg/lit)	88.18	88.18	0.003681
10.	Magnesium (mg/lit)	292.28	974.25	0.018853
11.	Free CO_2 (mg/lit)	173.8	1738.0	0.061318
12.	D.O. (mg/lit)	6.80	81.20	0.072276
13.	B.O.D. (mg/lit)	87.48	1458.0	0.099790
14.	C.O.D. (mg/lit)	60.0	600.0	0.052576
15.	Fluoride (mg/lit)	1.12	112.0	0.387815

Table 17.8: Parameter-wise Actual Value (V_n), Quality Rating (q_n) and Calculated Value of $W_n log_{10} q_n$ at Site No. VII for Pre-monsoon Period

Sl.No.	*Parameter*	*Actual Value (V_n)*	*Quality rating (q_n)*	*$W_n log_{10} q_n$*
1.	pH	7.61	61.0	0.042234
2.	Turbidity (NTU)	2.08	41.6	0.061282
3.	Conductivity (µS/cm)	0.599	197.0	1.447430
4.	Alkalinity (mg/lit)	205.0	205.0	0.004375
5.	T.S. (mg/lit)	694.0	138.8	0.000810
6.	T.D.S. (mg/lit)	482.0	96.4	0.000750
7.	Chloride (mg/lit)	88.04	44.02	0.001555
8.	Hardness (mg/lit)	156.0	156.0	0.004150
9.	Calcium (mg/lit)	35.27	35.27	0.002928
10.	Magnesium (mg/lit)	134.51	448.36	0.016727
11.	Free CO_2 (mg/lit)	92.4	924.0	0.056125
12.	D.O. (mg/lit)	3.08	120.02	0.078699
13.	B.O.D. (mg/lit)	9.72	162.0	0.069691
14.	C.O.D. (mg/lit)	60.0	600.0	0.052576
15.	Fluoride (mg/lit)	0.234	23.4	0.259124

Table 17.9: Parameter-wise Actual Value (V_n), Quality Rating (q_n) and Calculated Value of $W_n log_{10} q_n$ at Site No. VII for Pre-monsoon Period

Sl.No.	Parameter	Actual Value (V_n)	Quality rating (q_n)	$W_n log_{10} q_n$
1.	pH	7.49	49.0	0.039983
2.	Turbidity (NTU)	3.12	62.4	0.067947
3.	Conductivity (µS/cm)	2.05	683.0	1.788052
4.	Alkalinity (mg/lit)	445.0	445.0	0.005012
5.	T.S. (mg/lit)	1406.0	281.2	0.000926
6.	T.D.S. (mg/lit)	1250.0	250.0	0.000907
7.	Chloride (mg/lit)	268.38	134.19	0.002013
8.	Hardness (mg/lit)	352.0	352.0	0.004819
9.	Calcium (mg/lit)	141.08	141.08	0.004068
10.	Magnesium (mg/lit)	266.04	886.81	0.018595
11.	Free CO_2 (mg/lit)	242.0	2420.0	0.064038
12.	D.O. (mg/lit)	3.40	116.64	0.078231
13.	B.O.D. (mg/lit)	29.16	486.0	0.084741
14.	C.O.D. (mg/lit)	18.4	184.0	0.042861
15.	Fluoride (mg/lit)	0.210	21.0	0.250230

Table 17.10: Parameter-wise Actual Value (V_n), Quality Rating (q_n) and Calculated Value of $W_n log_{10} q_n$ at Site No. IX for Pre-monsoon Period

Sl.No.	Parameter	Actual Value (V_n)	Quality rating (q_n)	$W_n log_{10} q_n$
1.	pH	7.00	0.0	0.023657
2.	Turbidity (NTU)	2.44	48.8	0.004145
3.	Conductivity (µS/cm)	0.414	138.0	1.349911
4.	Alkalinity (mg/lit)	155.0	155.0	0.004145
5.	T.S. (mg/lit)	438.0	87.6	0.000735
6.	T.D.S. (mg/lit)	390.0	78.0	0.000716
7.	Chloride (mg/lit)	63.9	31.95	0.001423
8.	Hardness (mg/lit)	116.0	116.0	0.003906
9.	Calcium (mg/lit)	82.07	82.07	0.003622
10.	Magnesium (mg/lit)	95.96	319.86	0.015802
11.	Free CO_2 (mg/lit)	68.2	682.0	0.053629
12.	D.O. (mg/lit)	4.54	104.83	0.076475
13.	B.O.D. (mg/lit)	16.2	270.0	0.076689
14.	C.O.D. (mg/lit)	76.8	768.0	0.054605
15.	Fluoride (mg/lit)	0.184	18.4	0.239367

Table 17.11: Parameter-wise Actual Value (V_n), Quality Rating (q_n) and Calculated Value of $W_n \log_{10} q_n$ at Site No. X for Pre-monsoon Period

Sl.No.	*Parameter*	*Actual Value (V_n)*	*Quality rating (q_n)*	*$W_n log_{10} q_n$*
1.	pH	7.09	9.0	0.022573
2.	Turbidity (NTU)	2.60	52.0	0.064950
3.	Conductivity (μS/cm)	0.402	134.0	1.341852
4.	Alkalinity (mg/lit)	155.0	155.0	0.004145
5.	T.S. (mg/lit)	104.0	20.8	0.000498
6.	T.D.S. (mg/lit)	66.0	13.2	0.000424
7.	Chloride (mg/lit)	31.24	15.62	0.003022
8.	Hardness (mg/lit)	124.0	124.0	0.003961
9.	Calcium (mg/lit)	35.27	35.27	0.002928
10.	Magnesium (mg/lit)	102.51	341.70	0.015983
11.	Free CO_2 (mg/lit)	77.0	770.0	0.054626
12.	D.O. (mg/lit)	5.02	99.77	0.075662
13.	B.O.D. (mg/lit)	6.48	108.0	0.064137
14.	C.O.D. (mg/lit)	56.0	560.0	0.052009
15.	Fluoride (mg/lit)	0.210	21.0	0.250230

Table 17.12: Parameter-wise Actual Value (V_n), Quality Rating (q_n) and Calculated Value of $W_n \log_{10} q_n$ at Site No. XI for Pre-monsoon Period

Sl.No.	*Parameter*	*Actual Value (V_n)*	*Quality rating (q_n)*	*$W_n log_{10} q_n$*
1.	pH	6.88	12.0	0.025529
2.	Turbidity (NTU)	2.12	42.4	0.002331
3.	Conductivity (μS/cm)	0.551	183.0	1.427233
4.	Alkalinity (mg/lit)	175.0	175.0	0.004244
5.	T.S. (mg/lit)	692.0	138.4	0.000810
6.	T.D.S. (mg/lit)	628.0	125.6	0.000794
7.	Chloride (mg/lit)	46.86	23.43	0.001296
8.	Hardness (mg/lit)	154.0	154.0	0.004139
9.	Calcium (mg/lit)	24.05	24.048	0.002613
10.	Magnesium (mg/lit)	139.35	464.49	0.016824
11.	Free CO_2 (mg/lit)	50.6	506.0	0.051176
12.	D.O. (mg/lit)	3.56	114.95	0.077990
13.	B.O.D. (mg/lit)	12.96	216.0	0.073632
14.	C.O.D. (mg/lit)	82.4	824.0	0.055184
15.	Fluoride (mg/lit)	0.169	16.9	0.232378

Table 17.13: Parameter-wise Actual Value (V_n), Quality Rating (q_n) and Calculated Value of $W_n \log_{10} q_n$ at Site No. XII for Pre-monsoon Period

Sl.No.	Parameter	Actual Value (V_n)	Quality rating (q_n)	$W_n log_{10} q_n$
1.	pH	6.66	34.0	0.036229
2.	Turbidity (NTU)	2.32	46.4	0.063077
3.	Conductivity (µS/cm)	0.388	129.0	1.331434
4.	Alkalinity (mg/lit)	140.0	140.0	0.004061
5.	T.S. (mg/lit)	394.0	78.8	0.000717
6.	T.D.S. (mg/lit)	384.0	76.8	0.000713
7.	Chloride (mg/lit)	31.24	15.62	0.001129
8.	Hardness (mg/lit)	138.0	138.0	0.004049
9.	Calcium (mg/lit)	39.28	39.28	0.003016
10.	Magnesium (mg/lit)	114.07	380.23	0.016275
11.	Free CO_2 (mg/lit)	35.2	352.0	0.048193
12.	D.O. (mg/lit)	3.89	111.58	0.077501
13.	B.O.D. (mg/lit)	29.16	486.0	0.084741
14.	C.O.D. (mg/lit)	48.8	488.0	0.050878
15.	Fluoride (mg/lit)	0.164	16.4	0.229909

Table 17.14: Parameter-wise Actual Value (V_n), Quality Rating (q_n) and Calculated Value of $W_n \log_{10} q_n$ at Site No. I after the Onset of Monsoon

Sl.No.	Parameter	Actual Value (V_n)	Quality rating (q_n)	$W_n log_{10} q_n$
1.	pH	6.32	68.0	0.043350
2.	Turbidity (NTU)	2.46	49.2	0.064041
3.	Conductivity (µS/cm)	0.434	144.0	1.361571
4.	Alkalinity (mg/lit)	175.0	175.0	0.004244
5.	T.S. (mg/lit)	302.0	60.4	0.000674
6.	T.D.S. (mg/lit)	158.0	31.6	0.000567
7.	Chloride (mg/lit)	78.1	39.05	0.001506
8.	Hardness (mg/lit)	126.0	126.0	0.003974
9.	Calcium (mg/lit)	26.45	26.45	0.002692
10.	Magnesium (mg/lit)	109.88	366.27	0.016173
11.	Free CO_2 (mg/lit)	50.6	506.0	0.051176
12.	D.O. (mg/lit)	7.45	74.45	0.070850
13.	B.O.D. (mg/lit)	48.6	810.0	0.091738
14.	C.O.D. (mg/lit)	26.4	264.0	0.045828
15.	Fluoride (mg/lit)	0.201	20.1	0.246630

Table 17.15: Parameter-wise Actual Value (V_n), Quality Rating (q_n) and Calculated Value of $W_n \log_{10} q_n$ at Site No. II after the Onset of Monsoon

Sl.No.	Parameter	Actual Value (V_n)	Quality rating (q_n)	$W_n log_{10} q_n$
1.	pH	6.84	16.0	0.028485
2.	Turbidity (NTU)	2.44	48.8	0.063906
3.	Conductivity (μS/cm)	0.436	145.0	1.363467
4.	Alkalinity (mg/lit)	230.0	230.0	0.004469
5.	T.S. (mg/lit)	474.0	94.8	0.000748
6.	T.D.S. (mg/lit)	310.0	62.0	0.000678
7.	Chloride (mg/lit)	99.4	49.7	0.001605
8.	Hardness (mg/lit)	124.0	124.0	0.003961
9.	Calcium (mg/lit)	22.44	22.44	0.002556
10.	Magnesium (mg/lit)	110.33	367.75	0.016184
11.	Free CO_2 (mg/lit)	68.2	682.0	0.053629
12.	D.O. (mg/lit)	8.42	64.33	0.068448
13.	B.O.D. (mg/lit)	29.16	486.0	0.084741
14.	C.O.D. (mg/lit)	53.1	531.0	0.051572
15.	Fluoride (mg/lit)	0.268	26.8	0.270275

Table 17.16: Parameter-wise Actual Value (V_n), Quality Rating (q_n) and Calculated Value of $W_n \log_{10} q_n$ at Site No. III after the Onset of Monsoon

Sl.No.	Parameter	Actual Value (V_n)	Quality rating (q_n)	$W_n log_{10} q_n$
1.	pH	6.81	19.0	0.030250
2.	Turbidity (NTU)	2.83	56.6	0.066344
3.	Conductivity (μS/cm)	0.447	149.0	1.370922
4.	Alkalinity (mg/lit)	230.0	230.0	0.004469
5.	T.S. (mg/lit)	522.0	104.4	0.000764
6.	T.D.S. (mg/lit)	142.0	28.4	0.000550
7.	Chloride (mg/lit)	112.18	56.09	0.001654
8.	Hardness (mg/lit)	110.0	110.0	0.003863
9.	Calcium (mg/lit)	40.08	40.08	0.003033
10.	Magnesium (mg/lit)	85.58	285.26	0.015488
11.	Free CO_2 (mg/lit)	74.8	748.0	0.054388
12.	D.O. (mg/lit)	7.78	71.08	0.070088
13.	B.O.D. (mg/lit)	42.12	702.0	0.089778
14.	C.O.D. (mg/lit)	62.1	621.0	0.052859
15.	Fluoride (mg/lit)	0.204	20.4	0.247848

Table 17.17: Parameter-wise Actual Value (V_n), Quality Rating (q_n) and Calculated Value of $W_n log_{10} q_n$ at Site No. IV after the Onset of Monsoon

Sl.No.	Parameter	Actual Value (V_n)	Quality rating (q_n)	$W_n log_{10} q_n$
1.	pH	6.52	48.0	0.039771
2.	Turbidity (NTU)	2.82	56.4	0.066286
3.	Conductivity (μS/cm)	0.528	176.0	1.416548
4.	Alkalinity (mg/lit)	205.0	205.0	0.004375
5.	T.S. (mg/lit)	210.0	42.0	0.000614
6.	T.D.S. (mg/lit)	184.0	36.8	0.000592
7.	Chloride (mg/lit)	58.22	29.11	0.001385
8.	Hardness (mg/lit)	60.0	60.0	0.003365
9.	Calcium (mg/lit)	21.64	21.64	0.002527
10.	Magnesium (mg/lit)	46.81	156.04	0.013835
11.	Free CO_2 (mg/lit)	46.2	462.0	0.050428
12.	D.O. (mg/lit)	5.99	89.64	0.073902
13.	B.O.D. (mg/lit)	51.84	864.0	0.092622
14.	C.O.D. (mg/lit)	48.1	481.0	0.050759
15.	Fluoride (mg/lit)	0.227	22.7	0.256628

Table 17.18: Parameter-wise Actual Value (V_n), Quality Rating (q_n) and Calculated Value of $W_n log_{10} q_n$ at Site No. V after the Onset of Monsoon

Sl.No.	Parameter	Actual Value (V_n)	Quality rating (q_n)	$W_n log_{10} q_n$
1.	pH	6.69	31.0	0.035280
2.	Turbidity (NTU)	3.66	73.2	0.705711
3.	Conductivity (μS/cm)	1.96	653.0	1.775746
4.	Alkalinity (mg/lit)	790.0	790.0	0.005483
5.	T.S. (mg/lit)	1624.0	324.8	0.000950
6.	T.D.S. (mg/lit)	700.0	140.0	0.000812
7.	Chloride (mg/lit)	278.32	139.16	0.002028
8.	Hardness (mg/lit)	290.0	290.0	0.004660
9.	Calcium (mg/lit)	66.53	66.53	0.003450
10.	Magnesium (mg/lit)	243.46	83.15	0.018419
11.	Free CO_2 (mg/lit)	321.2	3212.0	0.047440
12.	D.O. (mg/lit)	5.02	99.77	0.075662
13.	B.O.D. (mg/lit)	77.76	1296.0	0.098177
14.	C.O.D. (mg/lit)	50.2	502.0	0.051110
15.	Fluoride (mg/lit)	0.262	26.2	0.268414

Table 17.19: Parameter-wise Actual Value (V_n), Quality Rating (q_n) and Calculated Value of $W_n \log_{10} q_n$ at Site No. VI after the Onset of Monsoon

Sl.No.	*Parameter*	*Actual Value (V_n)*	*Quality rating (q_n)*	*$W_n log_{10} q_n$*
1.	pH	7.31	31.0	0.035280
2.	Turbidity (NTU)	2.78	55.6	0.066051
3.	Conductivity (μS/cm)	1.98	660.0	1.778667
4.	Alkalinity (mg/lit)	210.0	210.0	0.004394
5.	T.S. (mg/lit)	344.0	68.8	0.000695
6.	T.D.S. (mg/lit)	324.0	64.8	0.000685
7.	Chloride (mg/lit)	112.18	56.09	0.001654
8.	Hardness (mg/lit)	100.0	100.0	0.003785
9.	Calcium (mg/lit)	40.88	40.881	0.003049
10.	Magnesium (mg/lit)	75.10	250.30	0.015130
11.	Free CO_2 (mg/lit)	59.4	544.0	0.052494
12.	D.O. (mg/lit)	6.32	86.27	0.073272
13.	B.O.D. (mg/lit)	55.08	918.0	0.093453
14.	C.O.D. (mg/lit)	63.3	633.0	0.053016
15.	Fluoride (mg/lit)	1.21	121.0	0.394168

Table 17.20: Parameter-wise Actual Value (V_n), Quality Rating (q_n) and Calculated Value of $W_n \log_{10} q_n$ at Site No. VII after the Onset of Monsoon

Sl.No.	*Parameter*	*Actual Value (V_n)*	*Quality rating (q_n)*	*$W_n log_{10} q_n$*
1.	pH	7.01	1.0	0.000000
2.	Turbidity (NTU)	2.02	40.4	0.060801
3.	Conductivity (μS/cm)	0.521	173.0	1.411838
4.	Alkalinity (mg/lit)	210.0	210.0	0.004394
5.	T.S. (mg/lit)	378.0	75.6	0.007110
6.	T.D.S. (mg/lit)	270.0	54.0	0.000655
7.	Chloride (mg/lit)	133.48	66.74	0.001726
8.	Hardness (mg/lit)	106.0	106.0	0.003832
9.	Calcium (mg/lit)	48.10	48.10	0.003183
10.	Magnesium (mg/lit)	76.70	255.65	0.015188
11.	Free CO_2 (mg/lit)	103.4	1034.0	0.057049
12.	D.O. (mg/lit)	7.61	72.77	0.070475
13.	B.O.D. (mg/lit)	51.84	864.0	0.092622
14.	C.O.D. (mg/lit)	72.1	721.0	0.054086
15.	Fluoride (mg/lit)	0.244	24.4	0.262564

Table 17.21: Parameter-wise Actual Value (V_n), Quality Rating (q_n) and Calculated Value of $W_n log_{10} q_n$ at Site No. VIII after the Onset of Monsoon

Sl.No.	*Parameter*	*Actual Value (V_n)*	*Quality rating (q_n)*	*$W_n log_{10} q_n$*
1.	pH	7.02	2.0	0.007121
2.	Turbidity (NTU)	3.16	63.2	0.068157
3.	Conductivity (μS/cm)	2.11	703.0	1.795959
4.	Alkalinity (mg/lit)	585.0	585.0	0.005236
5.	T.S. (mg/lit)	1176.0	235.0	0.000897
6.	T.D.S. (mg/lit)	1092.0	218.4	0.000885
7.	Chloride (mg/lit)	261.28	130.64	0.002002
8.	Hardness (mg/lit)	296.0	296.0	0.004676
9.	Calcium (mg/lit)	76.15	76.15	0.003561
10.	Magnesium (mg/lit)	249.60	832.01	0.018421
11.	Free CO_2 (mg/lit)	176.0	1760.0	0.061421
12.	D.O. (mg/lit)	5.67	93.02	0.074510
13.	B.O.D. (mg/lit)	113.4	1890.0	0.103345
14.	C.O.D. (mg/lit)	20.1	201.0	0.043588
15.	Fluoride (mg/lit)	0.231	23.1	0.258064

Table 17.22: Parameter-wise Actual Value (V_n), Quality Rating (q_n) and Calculated Value of $W_n log_{10} q_n$ at Site No. IX after the Onset of Monsoon

Sl.No.	*Parameter*	*Actual Value (V_n)*	*Quality rating (q_n)*	*$W_n log_{10} q_n$*
1.	pH	7.11	11.0	0.024635
2.	Turbidity (NTU)	2.34	46.8	0.063218
3.	Conductivity (μS/cm)	0.424	141.0	1.355803
4.	Alkalinity (mg/lit)	215.0	215.0	0.004414
5.	T.S. (mg/lit)	476.0	95.2	0.000748
6.	T.D.S. (mg/lit)	344.0	68.8	0.000695
7.	Chloride (mg/lit)	83.78	41.89	0.001534
8.	Hardness (mg/lit)	128.0	128.0	0.003987
9.	Calcium (mg/lit)	29.66	29.66	0.002786
10.	Magnesium (mg/lit)	109.93	366.43	0.016174
11.	Free CO_2 (mg/lit)	90.2	902.0	0.055927
12.	D.O. (mg/lit)	6.97	99.52	0.075621
13.	B.O.D. (mg/lit)	51.84	864.0	0.092622
14.	C.O.D. (mg/lit)	60.8	608.0	0.052685
15.	Fluoride (mg/lit)	0.189	18.9	0.241571

Table 17.23: Parameter-wise Actual Value (V_n), Quality Rating (q_n) and Calculated Value of $W_n log_{10} q_n$ at Site No. X after the Onset of Monsoon

Sl.No.	Parameter	Actual Value (V_n)	Quality rating (q_n)	$W_n log_{10} q_n$
1.	pH	7.04	4.0	0.014242
2.	Turbidity (NTU)	2.62	52.4	0.065076
3.	Conductivity (μS/cm)	0.401	133.0	0.957932
4.	Alkalinity (mg/lit)	230.0	230.0	0.004469
5.	T.S. (mg/lit)	354.0	70.8	0.000700
6.	T.D.S. (mg/lit)	136.0	27.2	0.000542
7.	Chloride (mg/lit)	79.52	39.76	0.001513
8.	Hardness (mg/lit)	116.0	116.0	0.003906
9.	Calcium (mg/lit)	31.26	31.26	0.002829
10.	Magnesium (mg/lit)	96.95	323.17	0.015830
11.	Free CO_2 (mg/lit)	118.8	1188.0	0.058191
12.	D.O. (mg/lit)	7.29	76.14	0.071219
13.	B.O.D. (mg/lit)	68.4	1134.0	0.096347
14.	C.O.D. (mg/lit)	50.2	502.0	0.051110
15.	Fluoride (mg/lit)	0.230	23.0	0.257707

Table 17.24: Parameter-wise Actual Value (V_n), Quality Rating (q_n) and Calculated Value of $W_n log_{10} q_n$ at Site No. XI after the Onset of Monsoon

Sl.No.	Parameter	Actual Value (V_n)	Quality rating (q_n)	$W_n log_{10} q_n$
1.	pH	6.89	11.0	0.021363
2.	Turbidity (NTU)	2.13	42.6	0.061673
3.	Conductivity (μS/cm)	0.562	187.0	1.433157
4.	Alkalinity (mg/lit)	190.0	190.0	0.004312
5.	T.S. (mg/lit)	428.0	85.6	0.000731
6.	T.D.S. (mg/lit)	332.0	66.4	0.000689
7.	Chloride (mg/lit)	69.58	34.79	0.001485
8.	Hardness (mg/lit)	94.0	94.0	0.003734
9.	Calcium (mg/lit)	42.48	42.48	0.003081
10.	Magnesium (mg/lit)	68.12	227.05	0.014863
11.	Free CO_2 (mg/lit)	70.4	704.0	0.053890
12.	D.O. (mg/lit)	8.59	62.64	0.068011
13.	B.O.D. (mg/lit)	100.44	1674.0	0.101682
14.	C.O.D. (mg/lit)	90.6	906.0	0.055963
15.	Fluoride (mg/lit)	0.177	17.7	0.236179

Table 17.25: Parameter-wise Actual Value (V_n), Quality Rating (q_n) and Calculated Value of $W_n \log_{10} q_n$ at Site No. XII after the Onset of Monsoon

Sl.No.	Parameter	Actual Value (V_n)	Quality rating (q_n)	$W_n \log_{10} q_n$
1.	pH	6.92	8.0	0.021363
2.	Turbidity (NTU)	2.44	48.8	0.063906
3.	Conductivity (uS/cm)	0.399	133.0	1.339800
4.	Alkalinity (mg/lit)	190.0	190.0	0.004312
5.	T.S. (mg/lit)	386.0	77.2	0.000714
6.	T.D.S. (mg/lit)	252.0	50.4	0.000644
7.	Chloride (mg/lit)	55.38	27.69	0.001364
8.	Hardness (mg/lit)	90.0	90.0	0.003698
9.	Calcium (mg/lit)	25.65	25.65	0.002666
10.	Magnesium (mg/lit)	74.37	247.9	0.015103
11.	Free CO_2 (mg/lit)	50.6	506.0	0.051176
12.	D.O. (mg/lit)	7.45	74.45	0.070850
13.	B.O.D. (mg/lit)	48.6	810.0	0.091738
14.	C.O.D. (mg/lit)	42.2	422.0	0.049684
15.	Fluoride (mg/lit)	0.179	17.9	0.237103

Table 17.26: Site-wise Calculated Values of Water Quality Index for Pre-monsoon Period and After the Onset of Monsoon

Site No.	Name of Site	Water Quality Index	
		Pre-monsoon	Onset of Monsoon
I	MDA Colony	105.41	101.16
II	Jigar Vihar Colony	109.43	103.45
III	Kali Ji Temple	103.38	102.87
IV	Upstream Nawabpura	116.40	118.48
V	Nawabpura Nullah	296.23	287.22
VI	Downstream Nawabpura	382.32	376.53
VII	Upstream Barbalan	106.85	109.43
VIII	Barbalan Nullah	283.42	280.45
IX	Downstream Barbalan	73.87	98.27
X	MBD-BLY Rail Bridge	90.57	39.96
XI	MBD-BLY Road Bridge	94.66	115.06
XII	LKO-DLH Bypass Bridge	89.52	89.98

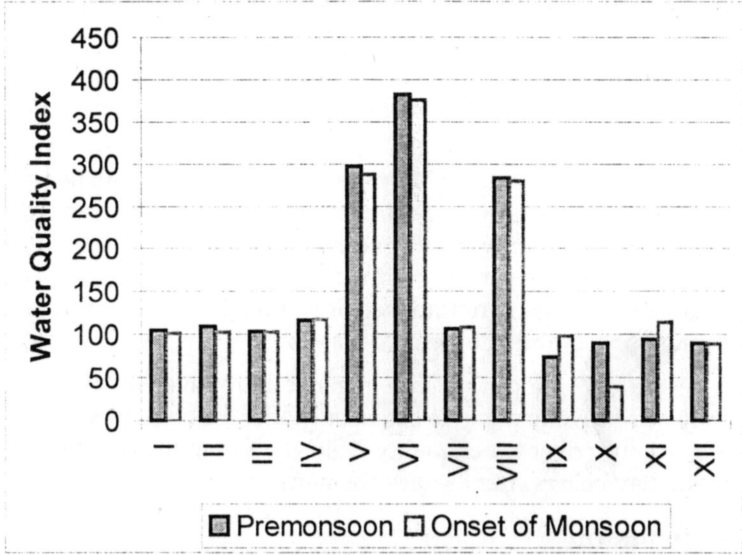

Figure 17.2: Graphical Presentation of Site-wise Seasonal Variation of Water Quality Index

Critical analysis of the data of the WQI presented in Table 17.26 reveals many interesting features regarding the status of Ram Ganga river water pollution at all the twelve different sampling sites at Moradabad during the course of study.

The values of Water Quality Index of Ram Ganga river water ranges from 73.87 to 382.32 for pre-monsoon period and from 39.96 to 376.53 after the onset of monsoon in the catchment area of our study. The values of WQI are found to be higher than 100 at all the sites for both the periods except at Site no. IX, X, XI and XII for pre-monsoon period and at Site no. IX, X and XII only after the onset of monsoon. Therefore, it can be said that the Ram Ganga river water is severely polluted at Site no. I, II, III, IV, VI and VII for both periods. An overlook of the data presented in Table 17.26 and its comparison with the standards of WQI for rest of the river sites indicates that river is moderately polluted at Site no. IX with 73.87 value of WQI and it is excessively polluted at Site no. X, XI and XII for pre-monsoon period. River water is excessively polluted at Site no. IX and XII and severely polluted at Site no. XI after the onset of monsoon. Surprisingly, river water at Site no. X becomes fit for even human consumption after the onset of monsoon though it was excessively polluted for pre-monsoon period.

A marked increase in the values of WQI in the downstream Ram Ganga river water at Nawabpura is noticed for both periods. Therefore, inference can be drawn that the level of river water pollution increases with the mixing up of Nawabpura nullah carrying mixed discharge of industrial effluents and city discharge. This phenomenon is not noticed at Site no. IX and this might be because of some other unknown reasons.

Figure 17.2 indicates that in general, the river water quality shows improvement but to extent only after the onset of monsoon. Remarkably, river water quality at Site no. X becomes suitable even for human beings after the onset of monsoon.

Conclusion

On the basis of above exhaustive discussions for Ram Ganga river water quality at Moradabad and on the basis of calculated values of Water Quality Index, one may conclude that the Ram Ganga river is severely polluted for pre-monsoon period as well as after the onset of monsoon. River water quality is improved before leaving the Moradabad city because of lesser human activities and absence of mixing up of effluents with river water. Mixing up of Nawabpura nullah carrying city discharge and industrial effluents is playing its role in multiplying the river water contamination. Another relevant conclusion that can be drawn from the present study is that the river water quality shows improvement after the onset of monsoon. River water is uncontaminated only at Site no. X after the onset of monsoon. Assessment of aquatic environment on the basis of Water Quality Index is proved to be an important and effective tool. Some strict and effective measures are urgently required to check the level of contamination of Ram Ganga river at Moradabad so that flora and fauna may also grow easily in the river water.

References

APHA, AWWA, WPCF, 1995. *Standard Methods for Examination of Water and Wastewater,* 19th edn. American Public Health Association, Washington, D.C. 2005, U.S.A.

Chaudhari, G.R., Sohani, Deepali and Srivastava, V.S., 2004, Groundwater quality index near industrial area. *Ind. J. Env. Prot.*, 24(1): 29–32.

Pradhan, S.K., Patnaik, D. and Rout, S.P., 2001, Water quality index for the ground water around a phosphatic fertilizer plant. *Ind. J. Env. Prot.*, 21(4): 355–358.

Singh, A.P. and Ghosh, S.K., 1999. Water quality index for river Yamuna. *Poll. Res.*, 18: 435–439.

Sinha, D.K., Saxena, Shilpi and Saxena, Ritesh, 2004. Water quality index for Ram Ganga river water at Moradabad. *Poll. Res.*, 23(3): 527–531.

Sinha, D.K. and Srivastava, A.K., 1994. Water quality index for river Sai at RaeBareli for the pre-monsoon period and after the onset of monsoon. *Ind. J. Env. Prot.*, 14(5): 340–345.

Sinha, A.K., 1989. Water quality index for river Ganga water between Shuklaganj and Kalakakar. In: *Management of Aquatic Ecosystem*, pp. 233–242.

Tiwari, T.N., Das, S.C. and Bose, P.K., 1986. Weighed geometric water quality index for river Jhelum in Kashmir. *Journal of M.A.C.T.*, 19: 33–41.

Tiwari, T.N. and Mishra, M., 1985. A preliminary assignment of water quality index of major Indian rivers. *Indian J. Env. Prot.*, 5(4): 276–279.

Chapter 18

Oviposition Deterrent Activities of *Azadirachta indica*, *Dalbergia sissoo* and *Tagetes erecta* Leaf Extract Against *Culex quinquefasciatus* Say

Navpreet Kaur Gill and Shivali Uttam

Department of Zoology, Punjabi University, Patiala – 147 002

ABSTRACT

The Leaf extracts of *Azadirachta indica, Dalbergia sissoo* and *Tagetes erecta* were tested under laboratory conditions for oviposition deterrent activities against the adult mosquito *Culex quinquefasciatus* 1 per cent leaf extract of all the three plant leaves were prepared in acetone as well as in petroleum ether alcohol (Pet ether). The acetone control and pet ether control were also run simultaneously. Acetone extracts of *A. Indica* and *D. Sissoo* have shown high per cent repellency as compared to the pet ether extracts whereas pet ether extracts of *T. erecta* has shown more repellency than acetone extracts.

Introduction

Insect-transmitted disease remains a major source of illness and death worldwide. Mosquitoes transmit disease to more than 700 million people annually (Taubes, 1997). Malaria alone kills 3 million people each year, including 1 child every 30 seconds (Shell, 1997). Although mosquito-borne diseases currently represent a greater health problem in tropical and subtropical climates, no part of the world is immune to this risk (Fradin and Day, 2002).

Control of such diseases is becoming increasingly difficult because of increasing resistance of mosquitoes to pesticides (Ranson *et al.*, 2001). An alternative approach for mosquito control is the use of natural products of plant origin. The botanical insecticides are generally pest specific, readily biodegradable and usually lack toxicity to higher animals (Bowers, 1992). A large number of plant extracts have been reported to have mosquitocidal or repellent activity against mosquito vectors, but very few plant products have shown practical utility for mosquito control. One plant species may possess substances with a wide range of activities, for example extracts from the neem tree *Azadirachta indica* showed antifeedant, antioviposition, repellent and growth regulating activities (Schmutterer, 1995). Similarly *Dalbergia sissoo* and *Tagetes erecta* have also shown the anti oviposition and repellent activities against various insects (Hetheley *et al.*, 1988, Ansari *et al.*, 2000). Plant products can be obtained either from the whole plant or from a specific plant part by extraction with different types of solvents such as aqueous, methanol, choloroform, acetone and pet ether etc., depending on the polarity of the phytochemicals.

Mosquito oviposition is a complicated process regulated by a number of endogenous and exogenous factors. Biotic and abiotic features determine the choice of oviposition sites and subsequently the breading habitat for the growth and development of the progeny. In an effort to explore the biological aspects of commercial plant products, other than their intended larvicidal capabilities, the current research was initiated to investigate the effects of emulsifiable concentration of different plant extracts *i.e. Azadirachta indica, Dalbergia sissoo* and *Tagetes erecta* on oviposition of the mosquito, *Culex quinquifasciatus* in the laboratory.

Material and Methods

Leaves of *Azadirachta indica, Dalbergia sissoo, Tagetes erecta* were collected from the campus of Punjabi University, Patiala. The leaves were washed with water, dried and powdered separately by using a mechanical grinder. Powdered leaves (1 Kg.) of each plants were extracted with acetone and pet ether (Petroleum ether) (3:1) in a soxlhlet apparatus for 8 hours at 50°C. Then the extraction was concentrated in a rotary vacuum evaporator to get the greenish material which was used for bioassays.

The larvae of *Culex quinquefasciatus* were reared in the laboratory following the method suggested by Ansari *et al.* (1978). The oviposition deterrent activity was performed by using the method of Xue *et al.* (2001). Fifteen gravid females of *Culex quinquifasciatus* were (10 days, 4 days after blood feeding) transferred to each mosquito cage (45 cm × 38 cm × 38 cm) covered with plastic screen, with a glass top and a muslin sleeve for access. A 10 per cent sucrose solution was available at all the times. 100 ml of the formulation of each extract were taken for further studies. Two plastic bowls were kept in a cage in opposite corners of each cage. One containing plant extract formulation and other containing control. Six replicates of the experiment were run for each extract. After 24 hours the number of eggs laid in treated and control bowls were observed. The percentage effective repellency for each leaf extract was calculated using the formula given by Raj Kumar and Jebanesan (2005).

$$ER\ (\%) = \frac{NC - NT}{NC} \times 100\ (\%)$$

Results and Discussion

Testing the efficacy of extracted components from different plants on a major house hold insect such as *Culex quinquefasciatus*, several plant extracts have been used against gravid females of *Culex quinquefasciatus*. During present studies 1 per cent extract of *Azadirachta indica, Dalbergia sissoo, Tagetes erecta* in acetone and pet ether alcohol along with control was used. 5 replicas of experiments were performed with each extract. These extracts showed significant effect on the oviposition behaviour of *Culex quinquefasciatus*.

It has been observed that percentage of eggs laid by females were considerably lower in water treated with plant extract than that of control medium. The results have been computed in Table 18.1 and Figures 18.1 and 18.2.

Table 18.1: Showing Comparative Efficacy of Plant Extracts on Oviposition Deterrent Activities Against *Culex quinquefasciatus*

Plants	*Concentration*	*%age of Egg Rafts Laid on Control Medium*		*%age of Egg Rafts Laid on Treated Medium*		*%age Repellency*	
		Acetone	*Pet ether*	*Acetone*	*Pet ether*	*Acetone*	*Pet ether*
Azadirachta indica	1 per cent	91.322	87.133	6.942	12.623	92.398	85.512
Dalbergia sissoo	1 per cent	77.120	71.213	24.231	27.274	68.580	61.700
Tagetes erecta	1 per cent	72.801	79.198	30.000	17.989	58.790	77.280

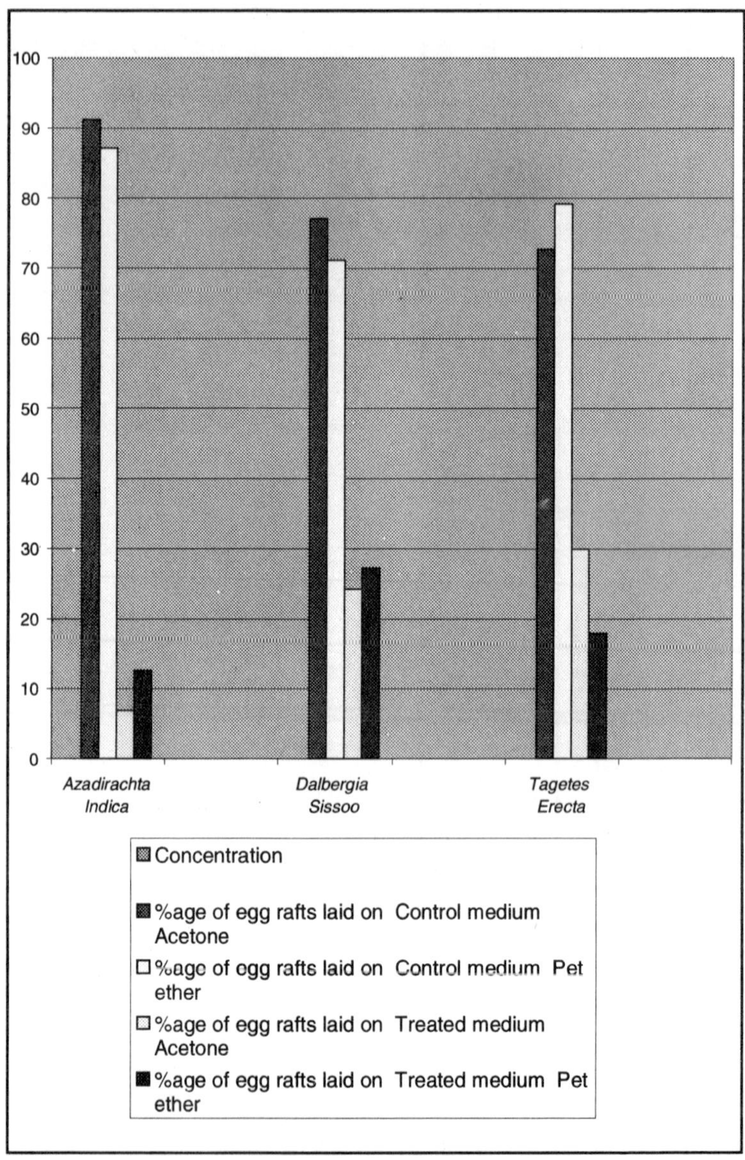

Figure 18.1: Showing Per cent Repellency of Plant Extracts on Oviposition Deterrent Activities Against *Culex quinquefasciatus*

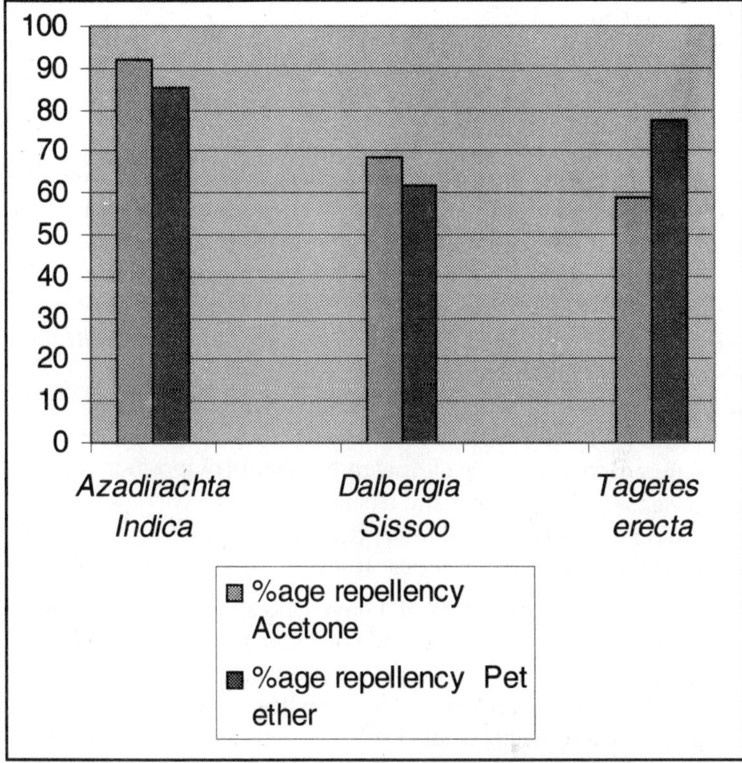

Figure 18.2: Showing Comparative Efficacy of Plant Extracts on Oviposition Deterrent Activities Against *Culex quinquefasciatus*

A. Indica acetone extracts exhibited the potential repellent action (92.398 per cent) whereas its pet ether counterpart also showed appreciable (85.512 per cent) repellency, containing 6.942 per cent and 12.623 per cent egg rafts on treated surface and 91.322 per cent and 87.133 per cent on control surface.

The acetone extract of *D. sisoo* showed 68.580 per cent repellency whereas its pet ether counter part showed 61.700 per cent repellency containing 24.231 per cent and 27.274 per cent of egg rafts laid on treated and 77.120 per cent and 71.213 per cent egg rafts laid on control surface.

The acetone extract of *T. erecta* showed 55.790 per cent repellency whereas its pet ether counter part showed supremacy with 77.280 per cent repellency 79.198 per cent and 72.80 per cent egg rafts were found on control surface and 30.000 per cent and 17.989 per cent egg rafts laid on treated surface.

It is therefore apparent that the extracts do not allow the females to settle over the water for egg laying. The females are devoid of egg laying opportunity, apparently because of the olfactory repellency covered by extracts. These observations are in conformation with the studies made by Yadav *et al.* (1987) who assessed disruption in oviposition behaviour of *Oligochaeta ramosa, Delonixrigia citrullus calocynthis* and *Calotropis procera* against *Anopheles stephensi, Culex quinquefasciatus* and *Aedes aegypti.* Present investigations are in direct conformity with the reports of Su and Mulla (1999), they reported that different concentration of *Azadirachta* disrupt oviposition of *Culex quinquefasciatus.* Raj Kumar and Jebanesan (2005) have reported that leaf extracts of *Solanum trilobatum* greatly reduce the number of eggs deposited by gravid *An. stephensi* at several concentration.

Dwivedi and Mathur (2001) have also asserted repellency of *Q. indica* and *Plutella xylostella* and concluded that acetone extracts proved more effective than the pet ether extracts. There observation support the results of present studies it has been observed that acetone extract of *Azadirachta Indica* and *Dalbergia sissoo* have more repellency than the pet ether extract. But in the case of *Tagetes erecta,* pet ether extract have shown supremacy over the acetone extract. Significant repellent properties of *Tagetes erecta* against mosquito species can be assigned for their oil contents. These results find support by the observation of Hethely *et al.* (1988) who isolated essential oils of *Tagetes erecta* and *T. patula,* which is used as insect repellent. Ansari *et al.* (2000) also isolated essential oil from *Dalbergia sissoo* which is used as repellent against *Anopheles stephensi, A. aegypti* and *Culex quinquefasciatus.* Llagas *et al.* (1995) observed efficacy of some local plant extract against different life stage of mosquito. Egg laying by gravid adults (oviposition repellent) was inhibited from extract of *Piper nigrum.* Similar results have been also reported during present investigation.

Hati *et al.* (1995) showed excellent repellent action of *Azadirachta indica* seed oil against *Aedes aegypti* mosquito. In general, published data have reported that neem products acts as oviposition repellents, deterrents, or inhibitors in a variety of test insects species. These insects include *Coleoptera,* such as the *bruchids,* *Callosobruchus maculatus* Fabricius, *C. analis* Fabricius and *C. chinensis* L. (Yadav, 1985); Lepidoptera, the Cabbage web warm, *Crocidolomia binotalis* Zeller (Fagoonee, 1981), the tobacco caterpillar, *Spodopttera litura* Fabricius (Joshi, 1987), and the Afro-Asian cotton bollworm, *Helicoverpa armigera* Hübner (Saxena and Rembold, 1984); and Diptera, the blowfly, *Lucilia cuprina* Wied. (Rice *et al.*, 1985) and the mosquitoes *Anopheles stephensi* Liston and *An. culicifacies* Giles (Dhar *et al.*, 1996), and *Culex quinquefasciatus* (Zebitz, 1987).

From the above discussion it has been concluded that the leaf extracts of all these plants exhibit oviposition deterrent activities against *Culex quinquefaciatus*. Further detailed studies are required to use these plants for vector control.

References

Ansari, M.A., Sharma, V.P. and Razdan, R.K., 1978. Mass rearing, procedures for *Anopheles stephensi* Liston. *J. Com. Dis.*, 10: 131–135.

Ansari, M.A., Razdan, R.K., Tondon and M. Vasudevan, P., 2000. Larvicidal and repellent action of *Dalbergia sissoo* Roxb. (F.) Leguminasae. oil against mosquito. *Biosource Technology*, 73(3): 207–211.

Bowers, W.S., 1992. Biorational approaches for insect control. *Korean Journal of Applied Entomology*, 31: 289–303.

Dhar, R., Dawar, H., Garg, S., Basir, S.F. and Talwar, G.P., 1996. Effects of volatiles from neem and other natural products on gonotrophic cycle and oviposition of *Anopheles stephensi* and *An. culicifacies*. *Journal of Medical Entomology* (In Press).

Dwivedi, S.C. and Mathur, M., 2001. Screening of repellent responses of IV instar larvae of diamond back moth, *Plutella xylostella* (L.) (Lepidoptera: Plutellidae) to five plant species. *J. Ecot. Res. Bicon.*, 2 (1 and 2): 65–69.

Fagoonee, I., 1981. Behavioral response of *Crocidolomia binotalis* to neem. In: *Proceedings of the First International Neem Conference*, (Rottach–Egern, Germany), (Eds.) H. Schmutterer, K.R.S. Ascher and H. Rembold. GTZ Eschborn, Germany, pp. 109–120.

Fradin, M.S. and Day, J.F., 2002. Comparative efficacy of insect repellents against mosquito bite. *New England Journal of Medicine*, 347(1): 13–18.

Hati, A.K., Bhawmik, K. Banerjee, A., Mukherjee, H., Poddar, G., Basu, D., Dhara, K.P. and Deya-Bhowmik, 1995. Repellent action of neem (*Azadirachta indica*), seed oil against *Aedes aegypti* mosquitoes. *Indian Journal of Dematology*, 40(4): 155–158.

Hethely, E. Tetenyi, P. Kaposi, P., Danos, B., Dernoczi, Z. and Budi, G., 1988. GC/MS investigation of antimicrobial and repellent compounds. *Herba Hungarica*, 27(2–3): 89–105.

Joshi, B.G., 1987. Use of need products in tobacco in India. In: *Proceedings of the Third International Neem Conference*, (Nairobi, Kenya), (Eds.) H. Schmutterer and K. R.S. Ascher. GTZ Eschborn, Germany, pp. 479–494.

Llagas, L., De-las and Cruz, I.R.F., 1995. Preliminary report on the efficacy of some botanical extracts against medically important mosquitoes in the Philippines. *Philippine Entomologist (Philippines)*, 9(2): 352.

Rajkumar, S. and Jobanesan, A., 2005. Oviposition deterrent and skin repellent activities of *Solanum trilobatum* leaf extract against the malarial vector *Anopheles Stephensi*. *Journal of Insect Science*, 5(15): 3.

Ranson, H., Rossiter, L., Ortelli, F., Jensen, B. Wang, X., Roth, C.W., Collins, F.H. and Hemingway, J., 2001. Identification of a novel class of insect glutathione S–transferases involved in resistance to DDT in the malaria vector *Anopheles gambiae*. *Biochemical Journal*, 359: 295–304.

Rice, M.J., Sextonm S. and Esmail, A.M., 1995. Antifeedant phytochemical blocks oviposition by sheep blowfly. *Journal of the Australian Entomological Society*, 24: 16.

Saxena, K.N. and Rembold, H., 1984. Orientation and ovipositional response of *Heliothis armigera* to certain neem constituents. In: *Proceedings of the Second International Neem Conference* (Rauischholzhausen, Germany), (Eds.) H. Schmutterer and K.R.S. Ascher. GTZ Eschborn, Germany, pp. 1999–210.

Schmutterer, H., 1995. *The Neem Tree Azadirachta indica and other Meliaceous Plants*. VCH Publishers, Weinneim, Germany, pp.

696.

Shell, E.R., 1997. Resurgence of a deadly diseases. *Atlantic Monthly*, August: 45–60.

Su, Tianyun and Mulla, Mir, S., 1999. Oviposition bioassay responses of *Culex tarsalis* and *Culex quinquefasciatus* to neem products containing azadirachtin. *Entomologia Experimentalis et Applicata*, 91(2): 337–345.

Taubes, G., 1997. A mosquito bites back. *New York Times Magazine*, August, 24: 20–46.

Xue, R.D., Barnard, D.R. and Ali, A., 2001. Laboratory and field evaluation of insect repellents as oviposition deterrents against the mosquito *Aedes albopictus*. *Medical and Veterinary Entomology*, 15: 126–131.

Yadav, R.S. Saxena, S.C. and Choubey, S.M., 1987. Disruption in oviposition behaviour of malaria, filariasis and dengue vectors by plant extracts. *Recent Trends in Ethology in India*, pp. 50–52.

Zebitz, C.P.W., 1987. Potential of neem seed kernel extracts in mosquito control. In: *Proceedings of the Third International Neem Conference* (Nairobi, Kenya), (Eds.) H. Schmutterer and K.Rr.S. Ascher. GTZ Eschborn, Germany, pp. 555–573.

Chapter 19

Histopathological and Histochemical Effects of a Newly Synthesized (JHA) Cyclohexloxy Compound on the Midgut on Fourth Instar Larvae of *Culex quinquefasciatus* Say

Navpreet Kaur Gill

Department of Zoology, Punjabi University, Patiala – 147 002

ABSTRACT

In Insects, Growth Regulators (IGRS) used for controlling insect pests includes Juvenile hormone analogue (JHA's), Moulting disruptors (MD's) and anti Juvenile hormones (AJH's). Juvenile hormone analogues are compounds which are as/or more potent than naturally occurring Juvenile hormones (JH). They act in the same way as the Juvenile hormone from Corpus allatum (Slama, 1971). During present studies, fourth instar larvae of pure colony were reared in 1 ppm, 2 ppm and 3 ppm doses of 1-(3'-methyl-6'-isopropy cyclohexy loxy 3,7-dimethyl-2(E)'6-octadiene for 24h. Histopathologically, all the three doses caused necrosis to midgut. The peritrophic membrane and brush broder were affected. Histochemically DNA, RNA, proteins, and 1:2 glycol groups, phospholipids and triglycerides showed a decrease where as glycogen showed an increase. It seems that JHA interferes in digestion and absorption of food and its metabolic products in the midgut of *Culex* resulting in mortality.

Introduction

The midgut in insects plays an important role in digestion as number of enzymes are secreted by it and also helps in absorption of digested food (Chapman, 1971). No extensive information is available concerning the nature of physiological effect of JHA on target species. It would prove helpful in identifying side effects of these compound on nontarget species. A very few workers, *viz.* Sharma (1994) and Mittal and Navpreet (1998) have reported effect of JHA's on midgut. Thus, the present work was undertaken to see the effect of JHA on the various metabolities of midgut epithelium of *Culex* with the help of histochemical techniques.

Materials and Methods

Fourth instar larvae (24h old) were collected from pure colony of *Culex quinquefasciatus* maintained at 26°C and relative humidity of 60-80 per cent in BOD incubator. 1 mg of newly sythesized JHA : 1-(3'-methyl-6'-isopropychclohexyloxy 3,7-dimethyl-2(E) 6-octadiene was dissolved in 10 ml of acetone and was diluted with distilled water to prepare sublethal doses of 1 ppm, 2 ppm and 3 ppm (calculated by Mittal and Navpreet 2000). Their controls were also prepared by dissolving 10 ml acetone in distilled water to prepare 1,2 and 3 ppm acetone. Fourth instar larvae were reared in these sublethal doses. These were processed and embedded for section cutting. The sections were stained with iron haematoxylin for histological study (Baker, 1945). Various histochemical tests, *viz.* mercuric bromphenol blue (Hg-BPB) and ninhydrin-Schiff for general proteins, periodic acid-Schiff (PAS) along with controls for 1:2 glycol groups, Best's carmine (BC) along with control for glycogen, methyl green/pyronin G along with controls for DNA and RNA, Sudan black B (SBB) for general lipids, acid haematein (AH) along with control for phospholipids, and Nile blue sulphate (NBS) for acidic and neutral lipids were performed according to the procedures detailed in Pearse (1968).

Results and Discussion

In control larvae, the midgut cells were observed to be flattened with central rounded nuclei. The nucleoli and cytoplasmic granules in iron haematoxylin were darkly stained. The peritrophic membrane covering the food was also observed (Figure 19.1). After treatment with 1 ppm the wall of the cells decreased in size. The cytoplasm of cells took very dark colour (Figure 19.2). After 3 ppm doses, width of the cells decreased. Nuclei, nucleoli and cytoplasm were very lightly stained (Figures 19.3 and 19.4). The peritrophic membrane and brush border showed necrosis, which was dose dependent.

DNA/RNA

In control larvae, the chromatin material of nuclei of midgut cells stained green in MG/PG test (–ve after control) and pink in Feulgen's technique (–ve after control) showing the presence of DNA in chromatin whereas the cytoplasmic granules near the apical surface showed more concentration and these stained pink in MG/PG for RNA (–ve after control for RNA). The brush border also stained pink showing presence of RNA. After treatement with 1 and 2 ppm doses, the depletion in staining in test for DNA and RNA showing abundant decrease in DNA and RNA in midgut cells as compared to control.

Proteins

In control larvae, the nucleus and nucleolus took very dark blue colour in Hg-BPB due to abundance of proteins. A rich deposition of protein being darkly stained in Hg-BPB was observed. Brush border and peritrophic membrane also stained light blue for moderate proteins in them. The pattern of staining in NHS for proteins was almost same. After treatment with 1 ppm and 2 ppm doses the nucleus, nucleolus and cytoplasmic granules showed lighter staining as compared to control. With 3 ppm dose there was abundant depletion in staining in Hg-BPB and NHS showing abundance decrease in proteins. Sharma (1994) also reported similar observations on midgut cells of *Culex* after 2 JHAs, *viz.* Isopropyl 12, 13-epithio-3, 9, 13-trimethyl-6-oxa-2 (E) (8E)-tetradecadienoate (JHA 1) and 2-(8, 12-dimethyl-5-oxa-7(E) 11-tridecadienyl)-P-ethyl phenyl either (JHA II). Mittal and Navpreet (1998a) reported severe damage to midgut epithelial cells of *Culex* after 1(3'-carbopropoxyphenoxy)-3, 7-dimethyl-6-octene. The necrosis on peritophic memrane (PM) serves two functions, (a) safeguardig the midgut epithelium against injuries by solid particles of food (Wigglesworth, 1972 a and b) selective ultrafiltration of digestive enzymes and end products of digestion (Zhuzhikov, 1964). According to Chapman (1971) if digestion is to occur in inesect, PM must be permeable to digestive enzymes and products of digestion. It seems that necrosis to PM and brush border by test compound may be responsible for decrease in absorbed food materials in midgut epithelial cells.

Carbohydrates

The cytoplasmic granules in control midgut epithelial cells stained purple for 1:2 glycol groups (–ve after PE and restored after KOH treatment). In BC, the glycogen granules were observed to be darkly stained. The intensity of colour increased with increase in

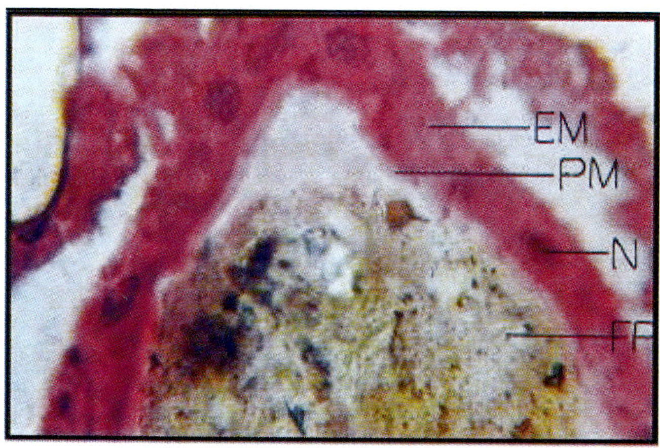

Figure 19.1
EM: Epithelial Membrance; PM: Peritrophic Membrance; N: Nucleus; FP: Food Particles

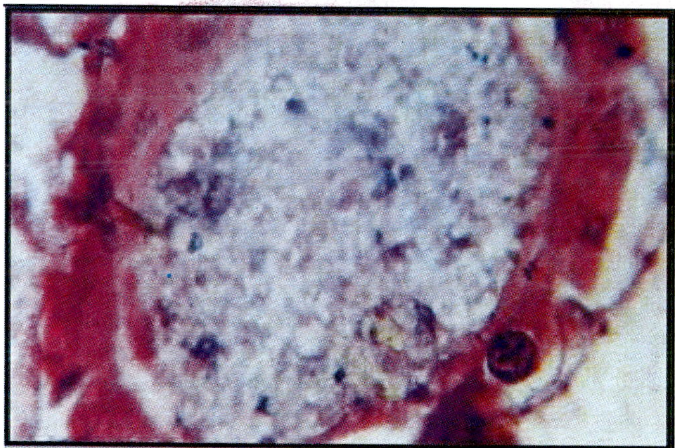

Figure 19.2

dose, *i.e.* 1 ppm, 2 ppm and 3 ppm respectively. In PAS, these cells took lighter colour as compared to control. Contrary to this Sharma (1994) reported a decrease in glycogen in midgut cells. Chapman (1971) reported that in *Aedes* larva glycogen appeared in the posterior midgut cells soon after glucose is ingested and the rapid conversion to glycogen might maintain the concentration gradient of glucose inwards from the lumen. This resulted in diffusion of glucose. It seems that test JHA interferes in this process.

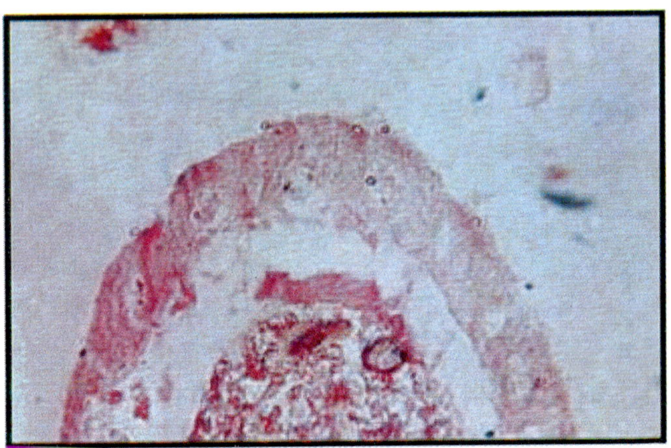

Figure 19.3

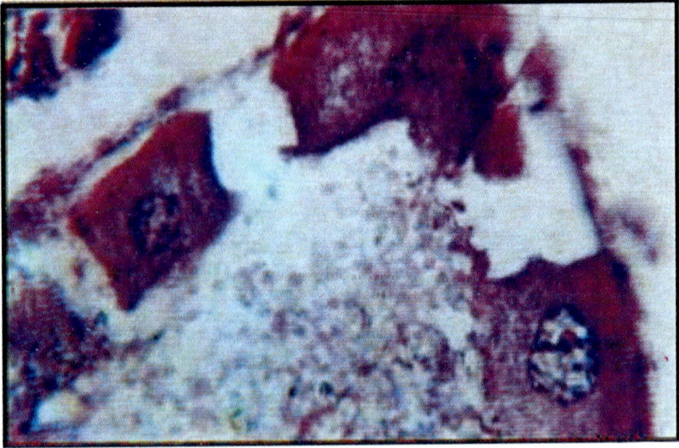

Figure 19.4

Lipids

In control midgut cells, the cytoplasmic stained dark blue in SBB for general lipids, blue in AH (–ve after control) for phospholipids. In NBS the deposition of pink coloured granules along the periphery of epithelium and brush border was observed. After treatment with 1 ppm and 2 ppm there was little change whereas with 3 pm lot of change in phospholipids and triglycerides was observed along with decrease in the size of cells. Triacyglycerols (TG) is the principal lipid class that is digested in the gut (Beenakkers *et al.*, 1985). Hydrolysis of TG is brought about by the action of lipases, Midgut lipase activity is reported in many insects, *viz. Periplanetra Americana* (Eisner, 1955), *Musca domestica* (House, 1974) and *Aedes aegypti* (Downer, 1978). It seems that JHA inhibits digestion of lipid. The decrease in total lipids, phospholipids, triglycerides in midgut epithelial cells has also been observed by Sharma (1994) in *Culex* after JHA I and JHA II treatment.

The gross depletion in metabolites such as lipids, carbohydrates and proteins in midgut epithelial cells after test compound may be responsible for reduced quantity of these in the fat body cells as reported by Sharma (1994) and Mittal and Navpreet (2000) may also cause depletion in whole larvae as reported by Mittal and Navpreet (1998b) and Mittal and Ruchita (1999).

References

Baker, J.R., 1945. *Cytological Technique*, 2nd edn. Methuen, London.

Beenakkers, A.M. Th. and van Marrewijk, I.E., 1985. Insect lipid and lipids and lipoproteins and their role in physiological processes. *Prog. Lipid Res.*, 24: 19–67.

Chapman, R.F., 1971. *The Insects: Structure and Function*. Hodder and Stoughton Ltd., England.

Downer, R.G.H., 1978. In: *Biochemistry of Insects*, (Ed.) M. Rockstein. Academic Press, New York, pp. 57–92.

Einser, J.J. 1955. *Exp. Zool.*, 130: 159–181 (Quoted from Beenakkers *et. al.*, 1985. House H.L. 1974. In: *Physiology of Insecta*, (Ed.) M. Rockstein. Academic Press, New York, 5: 63–117.

Mittal, P.K. and Navpreet, 1998a. Histological and cytochemical studies on the effect of JHA on body wall and midgut epithelium of *Culex pipiens quinquefasciatus* Say. *U.P.J. Zool.*, 18: 109–113.

Mittal, P.K. and Navpreet, 1998b. Biochemical effects of a newly synthesized chclohexyloxy compound JHA on the lipids of fourth instar larvae of *Culex pipens quinquefasciatus*. Say. *Curr. Sci.*, 74: 411–412.

Mittal, P.K. and Navpreet, 2000. Histological and histochemical studies on the fat body in Culex after treatment with a newly synthesized JHA. *U.P.J. Zool.*

Mittal, P.K. and Ruchita, 1999. Biochemical and TLC estimations of lipids of fourth instar of *Culex* and *Anopheles* treated with a newly synthesized geranly compound JHA. *J. Environ. Biol.*, 20(2): 149–152.

Pearse, A.G.E., 1968. *Histochemistry: Theoretical and Applied*. J. and A. Churchill, London.

Sharma, M., 1994. Bioassay, histopathological and biochemical effects of newly synthesized juvenile hormone, analogues on the mosquito, *Culex pipiens quinquefasciatus* Say. *Ph.D. Thesis*, Panjab University, Chandigarh.

Wigglesworth, V.B., 1972. *The Principles of Insect Physiology*. E.D. Button and Co. Inc., New York, p. 546.

Zhuzhikov, D.P., 1964. Function of the periitrophic membrance in *Musca domestica*, and *Calliphora erythrocephala* Meig. *J. Insect Physiol.*, 10: 273–278.

Chapter 20

Role of Some Abiotic Factors in Structuring the Macroinvertebrate Community with Special Reference to Insects Along a Longitudinal Profile of River Sindh, Kashmir Valley

Haroon Ul Rashid and Ashok K. Pandit

Aquatic Ecology Laboratory, P.G. Department of Environmental Science, The University of Kashmir, Srinagar – 190 006, J&K

ABSTRACT

The macroinvertebrate community structure, with special reference to insects, in relation to some abiotic factors (altitude, current velocity, water temperature and bottom substrate) has been studied in River Sindh of Kashmir valley. The sites along the stretch of the river, differing in these parameters, were observed to harbour totally different benthic macroinvertebrate communities. The insect larvae dominated, both in diversity and density, in swiftly flowing cold waters with boulder/cobble/pebble substrate rather than in slowly flowing waters with sand/clay (deposited) substrate.

Keywords: Macroinvertebrate community, Insects, Longitudinal profile, River Sindh, Kashmir valley.

Introduction

River Sindh, locally known as "Sendh", originates from the Panjtarni glacial fields in the Greater Himalaya at an elevation of more than 5,400 m (a.s.l). It falls an elevation of about 3433 m from its glacial sources upto Kangan in a distance of about 69 km, thus having a slope gradient of about 50 m in one kilometer (Kanth and Amin, 1996). Below Kangan the river drops only about 280 m upto Narayan Bagh in about 34 kilometers (Figure 20.1). The low slope gradient below Kangan results in the deposition of the sediments downstream. Sindh is a very important river of the J&K State as it has three hydro-electric power projects running with its waters. It inhabits many fish species especially the brown trout (*Salmo trutta* L.) for which the sport fishers come from across the globe for angling purposes.

Streams provide diverse habitats for different types of organisms including planktonic and periphytic algae, zooplankton, macroinvertebrates and fishes. The upstream processes are very important in studying the ecology of a river. Vannote *et al.* (1980) believe that the upstream processes directly influence the downstream ecology. Also such processes are important in determining the benthic communities of streams (Statzner and Higler, 1986; Vinson and Hawkins, 1998). Although Shiozawa (1983) believed that structuring of benthic macroinvertebrate community

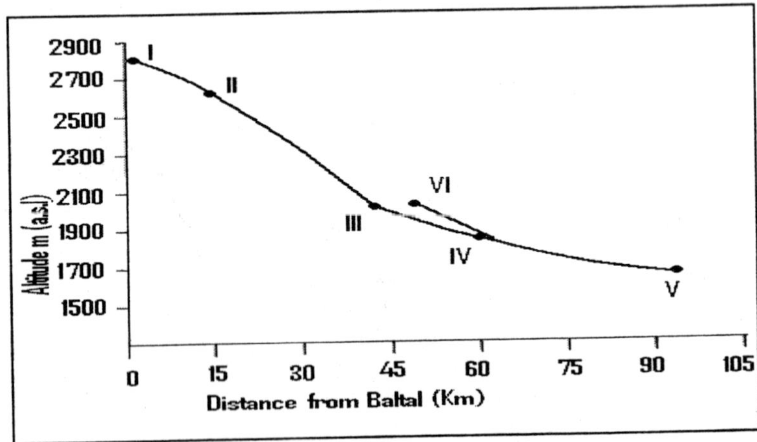

Figure 20.1: Longitudinal Profile of River Sindh Showing the Study Sites and Slope Gradient

is a function of both density dependent and density independent processes, yet abiotic habitat is generally considered as the major determinant of community organization in running waters (Hynes, 1970; Allan, 1995; Townsend *et al.*, 1997). It is in this backdrop that the present study was undertaken to assess the role of some abiotic variables in determining the structure of the macroinvertebrate community.

Study Area

River Sindh flows through the Sindh valley, a side valley of the main Kashmir valley, in north-east to south-west direction. It attains its shape at Baltal where two headwater streams (one coming from Zoji-la and the other coming from Panjtarni glacial fields) join to give birth to River Sindh which before merging with River Jhelum at Narayan Bagh (Shadipora) runs a distance of about 100 km (Figure 20.2). Wangat nalla, a major right bank tributary of Sindh, was also sampled at Wangat. The description of various study sites is given in Table 20.1.

Table 20.1: Description of Various Study Sites of River Sindh

Site	*Location*	*Altitude m (a.s.l)*	*Distance from Baltal (km)*	*River Bed Substrate**
I	Baltal	2810	2	Class A and B
II	Sonamarg	2625	17	Class A and B
III	Sumbal	2015	44	Class A and B
IV	Kangan	1860	61	Class A, B and C
V	Narayan Bagh	1580	95	Class C
VI	Wangat	2060	–	Class A and B

*: See material and methods.

The selection of the six sampling sites was made in such a way that the sites were least affected by the working of the three hydro-electric power projects as the sites fall well above the dams and below the power house exit canals.

Materials and Methods

The present study was undertaken in January (winter), April (spring), July (summer) and October (autumn) months of 2005. Among the physical parameters surface current velocity was

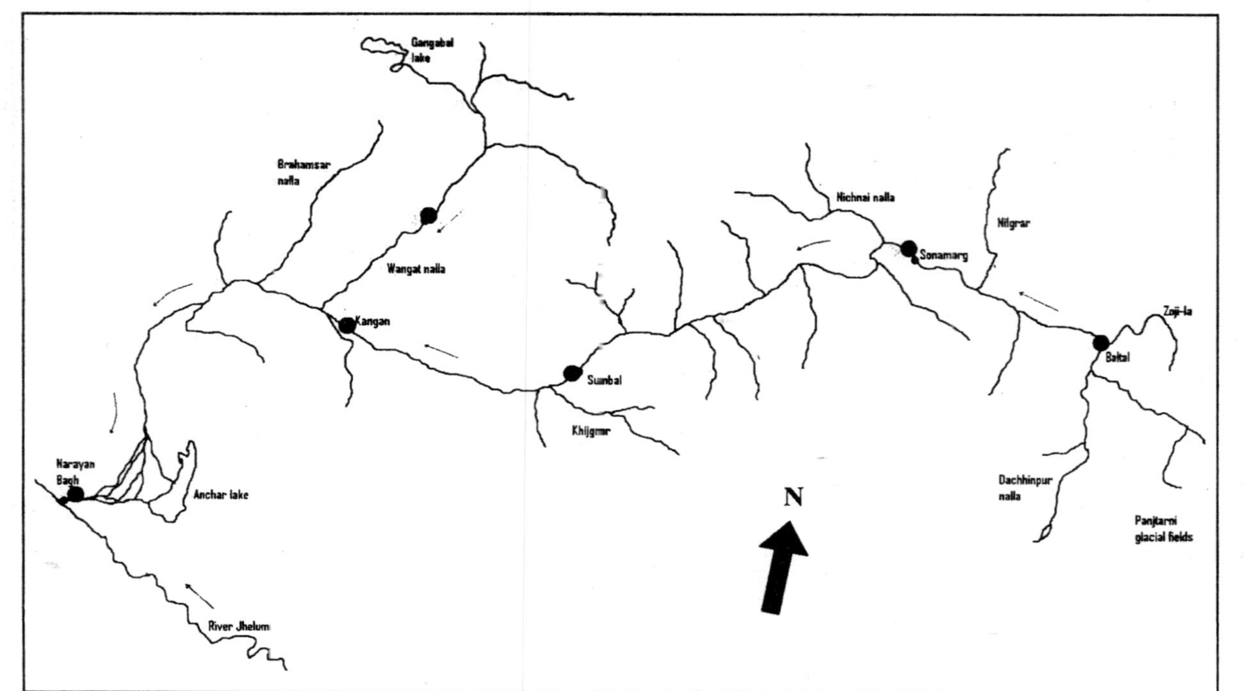

Figure 20.2: River Sindh Showing its Main Tributaries and the Study Sites

measured using a tennis ball which was let to flow for a stretch of 100m and the time taken by the ball to cover the distance was counted in order to calculate the velocity of water. Altitude was recorded with the help of Barigo Altimeter. Temperature was registered by using a mercury thermometer (–10 °C to 100 °C). The bottom substrates were classified visually in three classes viz, boulder/ cobble = Class A; pebble/gravel = Class B; and sand/clay (depositional) = Class C.

For benthic macroinvertebrate sampling, a Surber sampler (30cm × 30cm square quadrat) was used (Surber, 1936). The sampler was placed at proper positions facing upstream and then the bottom substrates were lifted and the sampler captured the drifted macroinvertebrates. The substrates lifted were placed in enamel trays where manual picking of undetached macroinvertebrates was done with the help of forceps. Ekman grab sampler (15 cm × 15 cm) was used at site V having soft bottom sediment (Ekman, 1911). The macroinvertebrate samples were preserved in 4 per cent formalin and/or 70 per cent alcohol (APHA, 1998). The identification of the preserved samples was done with the help of stereomicroscope using the standard keys given by Edmondson (1959), Pennak (1978), Engblom (1996), Engblom and Lingdell (1999) and Bouchard (2004).

Results and Discussion

In all 36 taxa of macroinvertebrates were encountered over entire stretch of the river. Of these, 30 belonged to Arthropoda (Insecta = 28, Crustacea = 1 and Hydracarina = 1) and 3 each to Annelida and Mollusca (Table 20.2). The distribution of the benthic macroinvertebrate fauna of River Sindh clearly depicted that there was a huge heterogeneity in the macroinvertebrate fauna of the sites differing in bottom substrate characteristics and the velocity of water (Tables 20.1 and 20.3). The Sindh, being an extremely turbulent river, brings down huge amounts of sediments which get deposited in the low slope gradient area (site V). The sites falling in the upper reaches of the river and having the bottom substrate of Class A and B were seen to have a greater diversity of the macroinvertebrate fauna, being dominated by insects as compared to the site V having low current velocity and soft and fine sediments, which harboured a greater density of molluscs and annelids (Table 20.4). The observation gains support from the findings of Downes *et al.* (2000) who found that the rough substrates were colonized by more taxa than fine substrates.

Table 20.2: List of Macroinvertebrate Taxa Collected from River Sindh

Class/Order/taxa
Arthropoda
Insecta
Diptera
Atherix sp.
Blepharocerid larva
Chironomous sp.
Diamesinae larva
Simulium sp.
Tabanidae
Tipulidae
Ephemeroptera
Acentrella sp.
Alainitis sp.
Baetis spp.
Caenis sp.
Drunella sp.
Ecdyonurus sp.
Epeorus sp.
Ephemerella sp.
Nigrobaetis sp.
Rhithrogena sp.
Trichoptera
Brachycentrus sp.
Glossosoma sp.
Hydropsyche sp.
Nectopsyche sp.
Rhyacophila sp.
Stenopsyche sp.
Plecoptera
Perlodidae

Contd...

Table 20.2–Contd...

		Coleopteran	
			Dytiscus larva
			Grouvellinus sp.
			Hydracanthus sp.
		Hemiptera	
			Gerris sp.
	Crustacean		
			Gammarus sp.
	Hydracarina		
			Hydracarina
Annelida			
			Erpobdella sp.
			Limnodrillus sp.
			Tubifex sp.
Mollusca			
			Coricula sp.
			Lymnaea sp.
			Radix sp.

Table 20.3: Physical Features of Water at Various Sites

Site	*Water Temperature (°C)*					*Velocity (cm/s)*				
	Jan.	*Apr.*	*Jul.*	*Oct.*	*Mean*	*Jan.*	*Apr.*	*Jul.*	*Oct.*	*Mean*
I	1.0	–	8.0	2.0	3.6	90	–	161	83	111.3
II	1.5	6.0	10.0	3.0	5.1	80	145	150	77	113.0
III	1.5	6.0	12.5	4.0	6.0	80	125	147	79	107.7
IV	4.0	10.5	12.5	4.0	7.7	85	154	163	71	118.2
V	5.0	9.5	15.5	9.0	9.7	28	55	77	33	48.2
VI	2.0	6.0	9.0	5.5	5.6	88	190	172	100	137.5

An examination of the data clearly revealed that on moving downstream, the per cent contribution of the insect larvae decreased but not to a greater extent until a totally different river bed substrate was encountered. The upper reaches of the river having high velocity of the water were observed to have maximum population of the insect larvae, thus giving rise to insect dominated macroinvertebrate community (Figure 20.3). This is attributed to the fact that the insect larvae have some well developed morphological adaptations to counter the stress exerted by the fast flowing water currents (Hynes, 1970). Further, Margalef (1960) emphasized the violence of flow as the main regulatory mechanism of benthic macroinvertebrate community in river systems. The lower values of standard deviation of the insect population further suggest that there is a stable community of benthic macroinvertebrates dominated by insects (Table 20.4). However, site V differed from other sites as regards the composition of the macrofaunal community, being dominated by molluscs and annelids. The insects contributed only about 1/4th to the total macroinvertebrate population at the site. The numerical surge of molluscs and annelids at site V is further related to the trophic conditions where detritus in the substrate constitutes the main source of food (Soszka, 1975; Pandit, 1980; Kaul *et al*, 1980; Pandit *et al.*, 1985).

Table 20.4: Population Density of Macroinvertebrates (/m^2), and Population Density (/m^2) of Insects (in parenthesis) at Different Sites of River Sindh

Site	*January*	*April*	*July*	*October*	*% Insect Fauna** (Mean Values)*
I	833 (833)	–	247 (242)	630 (577)	96.5 ± 3.26
II	546 (494)	460 (450)	315 (315)	651 (630)	96.2 ± 2.95
III	577 (556)	284 (274)	399 (378)	1365 (1293)	95.6 ± 0.85
IV	577 (514)	1179 (1147)	704 (693)	790 (751)	94.9 ± 2.97
V	176 (34)	248 (80)	285 (53)	179 (40)	23.0 ± 4.55
VI	1092 (995)	700 (643)	735 (662)	1171 (1107)	91.8 ± 1.30

**: Percent of bottom insect fauna to the total macroinvertebrate density in River Sindh.

Altitude and water temperature also play some role in the distribution of benthic macroinvertebrates as evinced by the distribution pattern of *Epeorus* sp. and *Hydropsyche* sp. The density of *Epeorus* sp. was very high at site I and II, low at site III and was altogether absent at site IV and downstream. Opposite was true for

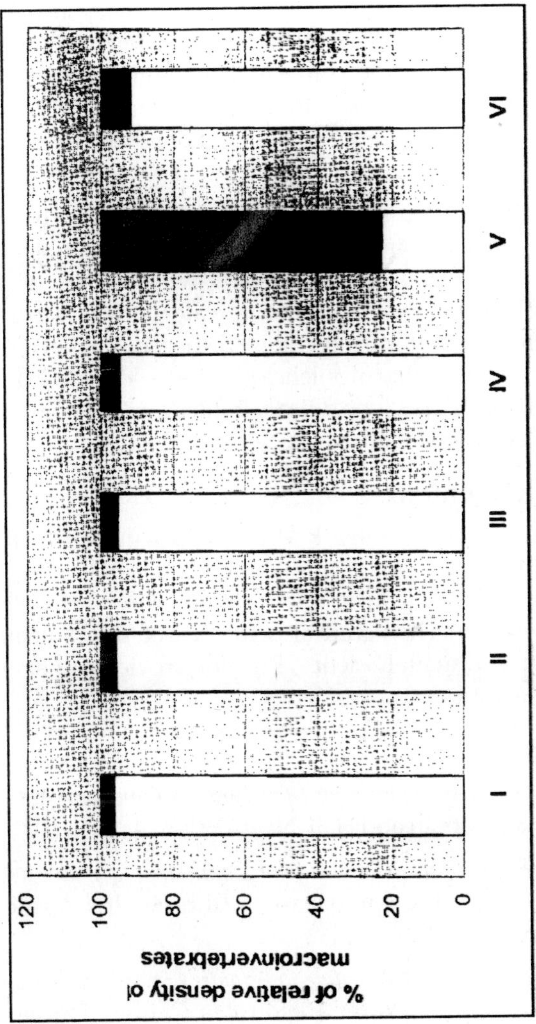

Figure 20.3: Relative Density of Insects and Other Macroinvertebrates

Hydropsyche sp. where density decreased while ascending upstream. This further lends credence to the view point held by Vannote *et al.* (1980), according to whom the stream benthic community structure is a function of characters such as altitude, thermal regime and stream size. In addition to the longitudinal characters, the localized physical characteristics are believed to produce distinct patterns in the structure of the benthic communities along stream continua (Grubaugh *et al.*, 1996). Most of the other insect forms were, however, found throughout the longitudinal gradient of the river excepting site V.

From the present study, it is inferred that the composition, growth and abundance of the macroinvertebrates in general and insects in particular is governed by a wide variety of physical characteristics of the river like altitude, water current velocity, water temperature and bottom substrate. Further, it is believed that the upstream processes like turbulence of water, sediment loading and detrital input and deposition directly influence the ecology of macro-invertebrates downstream.

Acknowledgement

The authors like to thank Mr. Gh. Hassan Rather, Research Scholar, CORD, K.U., Mr. Tahir Ahmed (uncle of the first author) and Mr. Tariq Ahmed (cousin of the first author) for their help during the collection trips. Thanks are also due to Prof. A.R. Yousuf, Head, Dept. of Environmental Science, K.U. for providing the necessary laboratory facilities.

References

Allan, J.D., 1995. *Stream Ecology: Structure and Functioning of Running Waters*. Chapman and Hall, New York.

APHA, 1998. *Standard Methods for Examination of Water and Waste Water*, 20th edn. American Public Health Association, Washington, D.C.

Bouchard, Jr., R.W., 2004. *Guide to Aquatic Macroinvertebrates of the Upper Midwest*. Water Resources Centre, University of Minnesota, St. Paul, Minnesota.

Downes, B.J., Lake, P.S., Schreiber, E.S.G. and Glaister, A., 2000. Habitat structure, resources and diversity: The separate effects of surface roughness and macroalgae on stream substrates. *Oecologia*, 123: 569–581.

Edmondson, W.T., 1959. *Freshwater Biology*. John Wiley, N.Y.

Ekman, S., 1911. Die bodenfauna des vättern, qualitativ and quantitativ untersucht. *Int. Revue ges. Hydrobiol. Hydrogr.*, 7: 146–204.

Engblom, E. and Lingdell, P.E., 1999. Analysis of benthic invertebrates. In: *River Jhelum, Kashmir Valley, Impacts on the Aquatic Environment*, (Ed.) Lennart Nyman. Swedmar, Sweden, p. 39–77.

Engblom, E., 1996. Ephemeroptera, Mayflies. In: *Aquatic Insects of North Europe: A Taxonomic Handbook*, (Ed.) A.N. Nilsson. Apollo Books, Denmark, p. 13–53.

Grubaugh, J.W., Wallace, J.B. and Houston, E.S., 1996. Longitudinal changes of macroinvertebrate communities along an Appalachian stream continuum. *Can. J. Fish. Aquat. Sci.*, 53: 896–909.

Hynes, H.B.N., 1970. *The Ecology of Running Waters*. Liverpool University Press. UK.

Kanth, T.A. and Amin, S.M., 1996. Drainage system in Kashmir valley. In: *Ecology, Environment and Energy*, (Eds.) A.H. Khan and A.K. Pandit. University of Kashmir, p. 185–197.

Kaul, V., Pandit, A.K. and Fotedar, D.N., 1980. Ecology of freshwater snails (Gastropod molluscs) in Haigam, a typical wetland of Kashmir. *Tropical Ecology*, 21: 32–46.

Margalef, R., 1960. Ideas for a synthetic approach to the ecology of running waters. *Int. Rev. ges. Hydrobiol.*, 45: 133–153.

Pandit, A.K., Pandit, S.N. and Kaul, V., 1985. Ecological relations between invertebrates and submerged macrophytes in two Himalayan lakes. *Poll. Res.*, 4: 53–58.

Pandit, A.K., 1980. Biotic factor and food chain structure in some typical wetlands of Kashmir. *Ph.D. Thesis*, The University of Kashmir, Srinagar.

Pennak, R.W., 1978. *Freshwater Invertebrates of United States*. John Wiley and Sons, London.

Shiozawa, D.K., 1983. Density independence versus density dependence in streams. In: *Stream Ecology: Application and Testing of General Ecological Theory*, (Eds.) J.R. Barnes and G.W. Minshall. Plenum Press, New York and London, p. 55–77.

Statzner, B and Higler, B., 1986. Stream hydraulics as major determinant of benthic invertebrate zonation patterns. *Freshwater Biology*, 16: 127–139.

Soszka, G.J., 1975. The invertebrates on submerged macrophytes in three Masurian lakes. *Ekol. Pol.*, 23: 371–391.

Surber, E.W., 1936. Rainbow trout and bottom fauna production in one mile of stream. *Trans. Am. Fish. Soc.*, 66: 193–202.

Townsend, C.R., Dolédec, S. and Scarsbrook, M.R., 1997. Species traits in relation to temporal and spatial heterogeneity in streams: a test of habitat templet theory. *Freshwater Biology*, 37: 367–387.

Vannote, R.L., Minshall, G.W., Cummins, K.W., Sedell, J.R. and Cushing, C.E., 1980. The river continuum concept. *Can. J. Fish. Aquat. Sci.*, 37: 130–137.

Vinson, M.R. and Hawkins, C.P., 1998. Biodiversity of stream insects: variation at local, basin and regional scales. *Ann. Rev. Entomol.*, 43: 271–293.

Chapter 21

Some Aspects of Habitat Ecology of Aquatic Entomofauna in two Freshwater Lakes of Kashmir Himalaya

Kawnsar-ul-Yaqoob and Ashok K. Pandit

Aquatic Ecology Laboratory, P.G. Department of Environmental Science, The University of Kashmir, Srinagar – 190 006, J&K, India

ABSTRACT

The present investigation deals with the habitat ecology of lacustrine insects of Dal and Nilnag lakes of Kashmir valley in relation to the depth of water column and the quality and quantity of aquatic macrophytes. Three main categories of aquatic insects, belonging to four different orders viz, Coleoptera, Hemiptera, Diptera and Odonata, have been recognized. It has been seen that the quality of water, the diversity and density of aquatic vegetation and suitable substratum are among the favourable factors increasing the potential of aquatic insects to inhabit their suitable ecological niches.

***Keywords**: Lacustrine, Substratum, Entomofauna, Macrophytes, Habitat, Ecology.*

Introduction

Biomonitoring is being much talked about and is a rapidly expanding field particularly in case of limnological studies as biomonitoring gives us an aggregate of all environmental stresses besides being a cost effective technique (Zajic, 1971). Aquatic invertebrates have served as an effective biological indicator for decades but amongst these entomofauna has been found to be a much more stable and reliable indicator of pollution. Though presently our knowledge of habitat ecology of aquatic insects in tropical waters of India is meager and is more so in high altitude Kashmir valley (Motwani *et al.*, 1956, Verma and Dalela, 1975; Ray *et al.*, 1986; Bisht and Das, 1979; 1980; Jan and Shah, 1985; Pandit *et al.*, 1985; Pandit, 1992, Shah and Pandit, 2001), yet considerable work in this regard has been done in the temperate regions (Gaufin and Tarzwell, 1952; Gaufin, 1956; Resh and Unzicker, 1976; Dennis and Patil, 1977).

Study Area

The present study is based on two lakes of Kashmir Dal and Nilnag. Dal lake with an area of 10.5 km^2, being a highly urban valley lake with greater anthropogenic pressures, is situated towards north-east of Srinagar city at an altitude of 1,586 m a.s.l (Figure 21.1). Nilnag lake with an area of 0.5 km^2, being a pine forest lake, is situated about 45 km in the south of Srinagar city at an altitude of 2,180 m a.s.l (Figure 21.2). The detailed description of these two lakes is already given by Pandit and Pandit (1996).

Material and Methods

The two lakes of Kashmir Dal (eutrophic) and Nilnag (mesotrophic) were surveyed at monthly intervals during 2004 from May to October for the habitat ecology of aquatic entomofauna. The insect fauna were collected from different depths with the help of cubic metal frame trap covered with net of mesh size–64 meshes/ cm^2 (Pandit 1980; Kaul *et al.*, 1980; Pandit and Kaul 1982). The trap was gently lowered in the macrophytic vegetation or sediments and lifted carefully with entire mass of macrophytes/sediments. The insect specimens collected were immediately preserved in 70 per cent alcohol with few drops of glycerine. Standard works of Edmondson (1963) and Pennak (1978) were used to facilitate the identification of the insect samples.

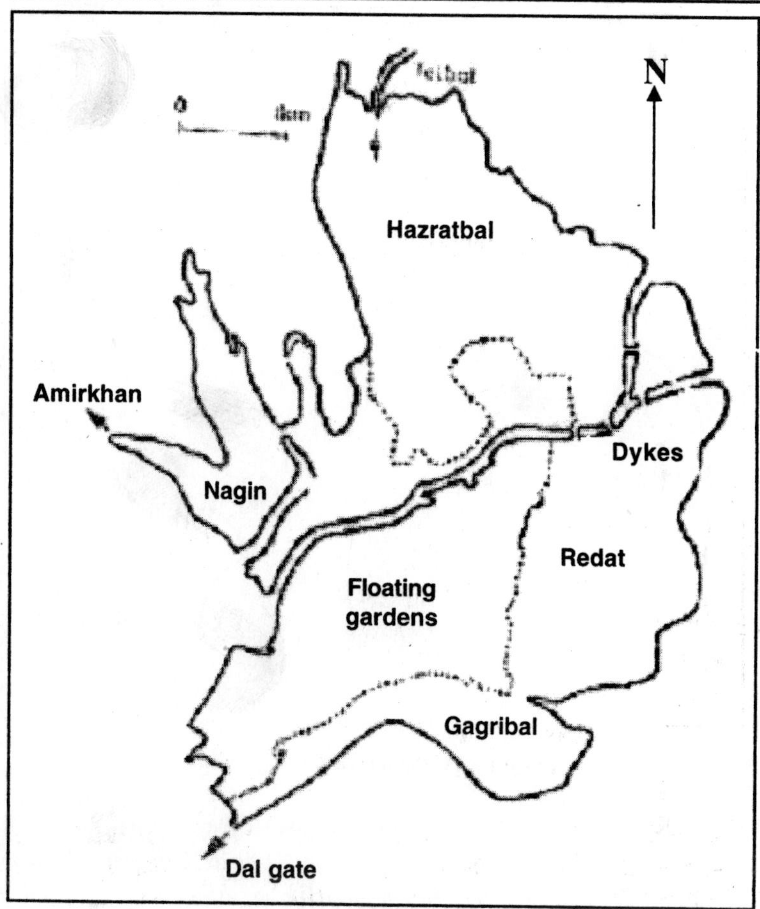

Figure 21.1: Map of Dal Lake

Results

The two Kashmir valley lakes were observed to differ radically in quality and quantity of macrophytes and the nature of the lake substratum. The common hydrophytes of Dal lake were *Phragmites australis, Typha angustata, Myriophyllum verticillatum, Nymphoides peltatum, Trapa natans, Hydrocharis dubia, Nymphaea* sp., *Ceratophyllum demersum, Hydrilla verticillata, Myriophyllum spicatum, Potamogeton*

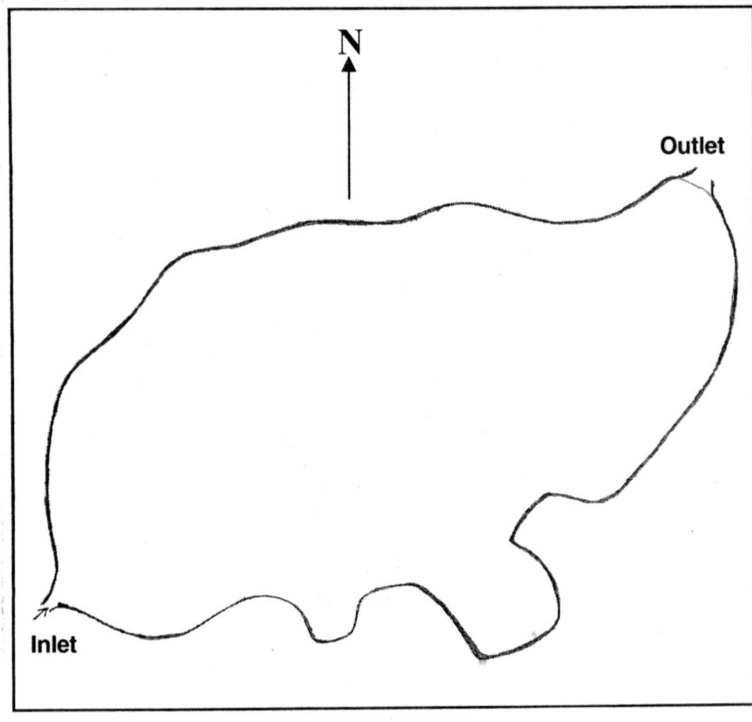

Figure 21.2: Map of Nilnag Lake

natans, P. crispus, P. lucens and P. pucillus, while in Nilnag in addition to these *Scirpus* sp., *Sagittaria sagittifolia, Alisma* sp., *Marsilea quadrifolia, Chara* sp., *Polygonum amphibium, Najas* sp., and *Potamogeton nodosus* were also recorded. The infestation of various macrophytes of the two lakes is, however, depicted in Table 21.1.

A perusal of data in Table 21.2 showed a total of 21 insect taxa, being collected from both the lakes, of which 18 were represented in Dal as against 12 being obtained from Nilnag lake. However, out of 21 genera only 10 were common to both the lakes. Lacustrine insects of Kashmir lakes were observed to belong to three main categories according to their ecological niches, mode of respiration and food and feeding habitats *viz,* (*i*) surface water community; (*ii*) mid-water macrophytic community and (*iii*) substratal community.

Table 21.1: Abundance of Aquatic Macrophytes in Dal and Nilnag Lakes

Sl.No.	*Macrophyte Species*	*Dal Lake*	*Nilnag Lake*
	A. Emergents		
	Tall growing emergents		
1.	*Phragmites australis*	++++	+++
2.	*Sparganium erectum*	++	++
3.	*Typha angustata*	++++	++
	Low growing emergents		
1.	*Sagittaria sagitifolia*	–	++
2.	*Myriophyllum verticillatum*	+++	+
3.	*Carex* sp.	+	+
	Ground layer species		
1.	*Alisma plantigina*	–	+
2.	*Marsilea quadrifolia*	–	+
3.	*Polygonum amphibium*	–	++
	B. Free floating types		
1.	*Chara* sp.	–	+
2.	*Lemna* spp.	+	+
3.	*Azolla azolla*	+++	–
4.	*Salvinia natans*	+	–
	C. Rooted floating–leaf types		
1.	*Nymphoides peltatum*	+++	+++
2.	*Trapa natans*	++	+++
3.	*Hydrocharis dubia*	++	++
4.	*Nelumbo nucifera*	+++	–
5.	*Nymphaea* sp.	+++	++
6.	*Potamogeton natans*	++	+
7.	*P. pectinatus*	++	–
	D. Submergeds		
1.	*Ceratophyllum demersum*	++++	+++
2.	*Hydrilla verticillata*	+++	+
3.	*Myriophyllum spicatum*	++++	+++
4.	*Najas graminae*	–	+
5.	*Potamogeton crispus*	+++	++
6.	*P. lucens*	+++	++
7.	*P. pucillus*	+	++
8.	*P. nodosus*	–	+

++++: Abundant; +++: Common; ++: Fairly Common; +: Rare, –: Absent.

Table 21.2: Prevalence of Different Insect in Dal and Nilnag Lakes

Sl.No.	*Order*	*Family*	*Dal*	*Nilnag*
		Coleoptera		
1.	*Dytiscus marginalis*	Dytiscideae	+	–
2.	*Hydrophilus* sp.	Hydrophilideae	+	+
3.	*Laccophilus* sp.	Dytiscideae	–	+
4.	*Dineutus* sp.	Gyrinidae		
		Hemiptera/Heteroptera		
1.	*Gerris* sp.	Gerridae	+	+
2.	*Notonectus undulata*	Notonectidae	+	+
3.	*Nepa* sp.	Nepidae	+	+
4.	*Sigara* sp.	Corixidae	–	+
		Diptera		
1.	*Pseudochironomus* sp.	Chironomidae	+	–
2.	*Chironomus* sp.	–do–	+	+
3.	*Pentaneura* sp.	–do–	+	+
4.	*Tendipes tentans*	–do–	+	–
5.	*Ablabesmyia* sp.	–do–	+	+
6.	*Chaoborus* sp.	Culicidae	+	–
7.	*Culex* sp.	–do–	+	+
8.	*Psychoda* sp.	Psychodidae	+	–
		Odonata		
1.	*Macromia* sp.	Macromidae	+	–
2.	*Aeshna* sp.	–do–	+	–
3.	*Coenagrion* sp.	–do–	+	+
4.	*Lestes* sp.	Lestidae	+	–
5.	*Helecordulia* sp.	Llbellulidae	+	+

+: Present; –: Absent.

Surface Dwelling Community

Insects, which pass most of their lives at the surface of the water and are supported by surface tension are included in this category. Two families *viz.* Gerridae and Gyrinidae are surface dwelling in Kashmir lakes studied. Gerridae occurs mostly in swarms but are found occasionally in groups of twos or threes. Among the Gerridae family, *Gerris* sp. is found in both the lakes (Table 21.2).

These surface dwelling bugs were observed to prefer the top waters above beds of macrophytes having little disturbances in the overlying water. The surfaced emergent vegetation and the little emergent parts of submerged macrophytes also provide them a place for food, shelter and hiding.

Among the coleopterans, *Dineutes* sp. have a similar habitat to that of surface dwelling bugs but are not represented in appreciable numbers. Unlike bugs these beetles when disturbed dive under the surface of water and rise shortly to the surface again.

Mid-water Macrophytic Community

These constitute the largest community of aquatic insects which pass most of their lives among aquatic vegetation, floating water, roots of marginal terrestrial weeds or even submerged parts of shrubs. They move up and down to obtain air and food and also escape from the danger of predation by harmonizing with the surroundings.

Among Coraxidae *Sigara* sp. in Nilnag lake have been seen to live among macrophytes and occasionally among floating vegetable wastes. They cling to the stems or leaves of macrophytes with the help of their long middle legs furnished with curved claws.

Among Notonectidae, *Notonectus undulata* (backswimmer) is found generally in groups and have a more or less similar habitat as that of coraxids. Among Nepidae family *Nepa* sp. was observed to inhabit mainly the hydrophytes in the shallow region with a muddy bottom. The great diving beetle, *Dytiscus marginalis* (a ferocious carnivore) was observed to inhabit the marginal vegetation of the lakes and used the macrophytes as place of egg laying and very rarely as a substrate. The noteworthy feature of Dal lake is the complete absence of *Laccophilus* sp. a representative of Dytiscidae family, indicating that the insect form avoids polluted biotopes.

Hydrophilus sp., belonging to Hydrophilidae family and reported from only Nilnag lake has been observed to prefer more specifically the macrophytes in the lake for purposes of egg laying and as food.

Among Odonata nearly all nymphal stages inhabit thick beds of macrophytes and floating plant debris, some of the genera collected were *Macromia, Coenagrion, Aeshna, Lestes* and *Helecordulia.*

With the exception of some species of Chironomidae which inhabit bottom organic debris, nearly all aquatic dipteran larvae were observed to be the inhabitants of aquatic macrophytes. The non-predatory Chironomidae uses macrophytes strongly as a mining place and as substrate and to some extent as a source of food (Pandit, 1984).

Substral Community

This category includes those aquatic insects, which pass most of their lives on the substratum of the lake. They may or may not come to the surface of the water to breathe fresh air depending on their mode of respiration. On the basis of the nature of the bottom of the lake these insects may be categorized into two types: (*a*) Saxicoline community and (*b*) Detrital community.

Saxicoline Community

In this category are included those insects which inhabit more preferably a rocky, stony or a pebbly and sandy substratum, without or with little amount of mud. It is interesting to note that as both the lakes have a muddy bottom with very little amount of sand, the aquatic insects adapted to saxicoline habitat could not be found in both the lakes.

Detrital Community

The second category of substratal community included some forms of Diptera, belonging to *Chironomus* sp., *Chaoborus* sp., *Pseudochironomus* sp., *Pentaneura* sp., *Tendipes tentans*, and *Ablabesmyia* sp. which were observed to prefer muddy substratum with sufficient organic matter as ooze. This category showed marked predominance in Dal lake as compared to Nilnag lake; thereby depicting high organic load in the Dal lake ecosystem.

Discussion

On the basis of present study regarding habitat ecology of aquatic insects, it is evident that the quality and quantity of aquatic vegetation and the nature of the lake bottom plays an important role in determining the distribution, diversity and abundance of insect communities. The differences in the intensity and diversity of macrophytic species in the two lakes, Dal and Nilnag, were the main cause of variation in the entomofauna sustained within the two different biotopes (Table 1). Needham (1938) and Rosin (1955) have also pointed out that with the increase and diversification of aquatic plants the productivity and number of insects per unit area also increases, an assumption also held by Frost (1942) and Macan (1963). Further, Usinger (1968) recognized *Chara* as inhabitant of less polluted waters, while Miller (1956) considered *Potamogeton* as an eutrophic species depicting that both the lakes are facing racing cultural eutrophication although the density of *Potamogeton* sp. is high in Dal lake as compared to Nilnag, the latter undergoing little trophic evolution. Conclusively, Pandit (1980) believes that macrophytes form the life environment of invertebrates.

The ecological relations between insects and macrophytes are reciprocal and manifold. The fauna uses the macrophytes directly as a place for living or breeding or even as food. Nevertheless the macrophytes in turn are damaged mainly due to grazing by the fauna sustained, though it is by no means easy to compare the composition of the number of individuals with the size of losses of macrophytes because of divergence between the place of feeding and living or breeding and the unknown time to bring about these losses. However, the damage caused to macrophytes is quite severe (Pandit *et al.*, 1985). Quite significant is also the indirect effect such a change in the environment as a result of different life processes of macrophytes and invertebrates. The complicated character of these relations have been pointed out by McGaha (1952), Gurzada (1959), Gaevskaja (1966), Krull (1970), Soszka (1975a,b), Kaul *et al.* (1980), Pandit (1980, 99) and Pandit *et al.* (1985). The community relation of entomofauna with hydrophytes is due to several advantages occurring to the insects *viz.* anchorage, natural hiding places, breeding sites, to check the rapid disturbances of water, to provide more oxygen and food spectrum and to afford suitable spawning niches, has also been highlighted by Bisht and Das (1979) and Pandit *et al.* (1985).

It is significant to note that Usinger (1968) reported the American species of *Dineutes* as typical stream inhabitant, which is probably one of the reasons for the rarity of this very Coleoptera in Kashmir lakes. Nayyar *et al.* (1976) stated hydrophilids preferably inhabit decomposing vegetable matter, but in Kashmir lakes they are characteristic live hydrophyte dwellers rather than inhabitants of decomposing vegetable material. Nayyar's generalization might be made from the fact that in general hydrophilids are natural scavengers of the mid-water macrophytes. It was further observed that the macrophytes provided an ideal resting place for the aquatic insect to move up and down the water column in search of food and to protect itself. However, Pandit (1992) reported *Hydrophilus* sp. (water scavenger beetles) to be largely bottom or sediment feeders having herbivorous/detritivorous type of feeding.

In addition to macrophytes the nature of lake bottom and the quality of water (polluted or unpolluted) generally affect the distribution of species of aquatic plant and the insects in a given environmental conditions. It has been observed that in Nilnag lake some species viz, *Sigara* sp., and *Laccophilus* sp. require special habitats which is not available in polluted lakes like the Dal. The same very reason explains the rarity of these insects in Dal lake. On the contrary, some species like *Chironomus* sp., *Chaoborus* sp., *Pentaneura* sp., *Pseudochironomus* sp., *Ablebesmyia* sp., *Tendipes tentans*, *Psychoda* sp. *Aeshna* sp. and *Macromia* sp., showing preference for eutrophic biotopes, have high population densities in Dal lake. The mere presence of some of these species in Nilnag lake may also indicate the gradual tropic evolution of the latter. Kaul and Pandit (1980) further believe that *Chironomus* sp. is basically a detritus feeder involved in detritus food chain of highly eutrophic wetlands of Kashmir Valley. This supports the present study that the preferred substrate for web spinning and case building chironomids is mud and organic ooze. Vass *et al.* (1977) also recognized red *Chironomus* as pollution indicators in Dal lake.

Acknowledgements

The authors are highly thankful to Head, Deptt. of Environmental Science for providing the necessary laboratory facilities during the course of the study. Thanks are also due to Mr. J.A. Javeed for his help in the field study.

References

Bisht, R.S. and Das, S.M., 1979. Studies on the ecology of aquatic entomofauna of Kumaon lakes. In: *Proc. Workshop on High Altitude Entomology and Wildlife Ecology*, Zoological Survey of India, Solan.

Dennis, B. and Patil, G.P., 1977. The use of community diversity indices for monitoring trends in water pollution impacts. *Tropical Ecology,* 19: 36–56.

Edmondson, W.T., 1959. *Freshwater Biology*. John Wiley and Sons Inc., New York, London.

Frost, S. W., 1942. *General Entomology*. Mc-Graw Hill Book Co. New York, London.

Gaevskaja, N.S., 1966. Rol. Vyssvysh Vodnych rastenij V. pitanii Zictncyh presyuch Vodoemov. *Izadaten Stwo*. Naka Moskova. 327 p.

Gaufin, A.R., 1956. Aquatic macro-invertebrates as indicators of organic pollution in Lytle Creek. *Sewage and Industrial Wastes,* 28: 906–924.

Gaufin, A.R. and Tarzwell, C.M., 1952. Aquatic invertebrates as indicators of stream pollution. *Publ. Hlth. Rep.*, 67: 57–64.

Gurzeda, A., 1959. Stosunki ekologizne miedzy fauna bezkregewa Goclinnesua zanurgone. *Ekol. Pol.*, 5: 139–146.

Jan, U. and Shah, G.M. 2002. Effect of physico-chemical factors on the distribution and abundance of macrozoobenthos in Dal lake, Kashmir. *Oriental Science,* 7(1): 35–40.

Kaul, V. and Pandit, A.K., 1980. Management of wetland ecosystems as wildlife habitats in Kashmir. In: *Proceedings of an International Seminar on Management of Environment,* (Ed.) B. Patel. Bhabha Atomic Research Centre, Mumbai, India, p. 31–52.

Kaul, V., Pandit, A.K. and Fotedar, D.N., 1980. Ecology of freshwater snails (gastropod molluscs) in Haigam, a typical wetland of Kashmir. *Tropical Ecology,* 21(1): 32–46.

Kaul, V. and Pandit, A.K., 1981. Benthic communities as indicators of pollution with reference to wetland ecosystems of Kashmir. In: *Proceedings of the WHO Workshop on Biological Indicators and Indices of Environmental Pollution,* (Eds.) A.R. Zafar, K.R. Khan, M.A. Khan and G. Seenayya. Cent. Bd. Prev. Cont. Water Poll., Osmania University, Hyderabad, India, p. 33–52.

Krull, J.L., 1970. Aquatic plant-macro-invertebrates associations and waterfowl. *J. Wildl. Manage.*, 34 (4): 707–718.

Macan, T.T., 1963. *Freshwater Ecology*. Longmans Co. Ltd., London.

McGaha, Y.J., 1952. The limnological relations of insects to certain aquatic flowering plants. *Trans. Am. Microscop. Soc.*, 71(4): 355–371.

Miller, L., 1956. Die Besiedlung der Potamogeton zone ostholsteinischer seen. *Arch. Hydrobiol.*, 52: 470–606.

Motwani, M.P., Banerjee, S. and Karamchandani, S.J., 1956. Some observations on the pollution of the river Sons by the factory effluents of the Rohtas Industries at Dalmia Nagar (Bihar). *Indian J. Fish.*, 3: 3–34.

Nayyar, K.K., Annanthakrishnan, T. N. and David, B.V., 1976. *General and Applied Entomology*. Tata Mc-Graw Hill Publishing Co. Ltd., New Delhi.

Needham, J.G., 1957. *A Guide to the Study of Freshwater Biology with Special Reference to Aquatic Insects and Other Invertebrate Animals and Phytoplankton*. Comstock Publishing Association, New York.

Pandit, A. K., 1980. Biotic factor and food chain structure in some typical wetlands of Kashmir. *Ph.D Thesis*, Kashmir University, Srinagar.

Pandit, A. K. and V. Kaul 1982. Trophic structure of some typical wetlands. In: *Wetlands: Ecology and Management, Part II*, (Eds.) B. Gopal, R.E.Turner, R.G. Wetzel and D.F. Whigham. Nat. Inst. Ecol. and Int. Sci. Publi., Jaipur, India, p. 55–82.

Pandit, A.K., 1984. Role of macrophytes in aquatic ecosystems and management of freshwater resources. *Journal of Environmental Management*, (London), 18: 73–88.

Pandit, A.K., Pandit, S.N. and Kaul, V., 1985. Ecological relations between invertebrates and submerged macrophytes in two Himalayan lakes. *Pollution Research*, 4(2): 53–58.

Pandit, A.K., 1992. Ecology of insect community in some typical wetlands of Kashmir Himalaya. In: *Current Trends in Fish and Fishery Biology and Aquatic Ecology*, (Eds.) A.R. Yousuf, M.K. Raina and M.Y. Qadri. P.G. Deptt. of Zoology, University of Kashmir, Srinagar, J&K, India, p. 271–286.

Pandit, A.K. and Pandit, S.N., 1996. Ecological studies of periphyton in two lakes of Kashmir Himalaya. In: *Assessment of Water*

Pollution, (Ed.) S.R. Mishra. APH Publishing Corporation, New Delhi, India, p. 175–199.

Pandit, A.K., 1999. *Freshwater Ecosystems of the Himalaya*. Parthenon Publishing, New York, London.

Pennak, R.W., 1978. *Freshwater Invertebrates of United States*. John Wiley and Sons, New York.

Resh, V.H. and Unzicker, J.D., 1975. Water quality monitoring and aquatic organisms: the importance of species identification. *J. Wat. Poll. Con. Fed.*, 47: 9–19.

Rosine, W.N., 1955. The distribution of invertebrates on submerged aquatic plant surfaces in Muskee lake, Colarado. *Ecology*,36: 308–314.

Roy, S.P., Kumar, V. and Pathak, H.S., 1986. Role of certain aquatic insects in the evaluation of water quality of a fish pond. *Biol. Bull. of India*, 8(2): 95–99.

Shah, K.A. and Pandit, A.K., 2001. Macro-invertebrates associated with macrophytes in various freshwater bodies of Kashmir. *J. Res. and Dev.*, 1: 44–54.

Soszka, G.J., 1975a. The invertebrates on submerged macrophytes in three Masurian lakes. *Ecol. Poll.*, 23(3): 371–391.

Soszka, G.J., 1975b. Ecological relation between invertebrates and submerged littoral. *Ecol. Poll.*, 23(3): 393–415.

Usinger, R.L., 1968. *Aquatic Insects of California*. California Univ. Berkley and Los. Angles, 3rd Reprint, 508 p.

Vass, K.K., Sunder, S. and Langer, R.K., 1977. Pollution indicators of Dal lake, Kashmir. *Proc. Ind. Sci. Cong.*, Part III, 183 p.

Verma, S.R. and Dalela, R.C., 1975. Studies on the pollution of the Kalinadi by industrial wastes near Mansurpur. Part II: Biological index of pollution and the biological characteristics of the river. *Acta Hydrochem. Hydrobiol.*, 3: 274–295.

Zajic, J.E., 1971. *Water Pollution: Disposal and Reuse*, Vol. 1. Marcel Dekker Inc., New York, 389 p.

Index